# Synthesis Lectures on Data Mining and Knowledge Discovery

**Series Editors**

Jiawei Han, University of Illinois at Urbana-Champaign, Urbana, USA

Lise Getoor, University of California, Santa Cruz, USA

Johannes Gehrke, Microsoft Corporation, Redmond, USA

The series focuses on topics pertaining to data mining, web mining, text mining, and knowledge discovery, including tutorials and case studies. Potential topics include: data mining algorithms, innovative data mining applications, data mining systems, mining text, web and semi-structured data, high performance and parallel/distributed data mining, data mining standards, data mining and knowledge discovery framework and process, data mining foundations, mining data streams and sensor data, mining multi-media data, mining social networks and graph data, mining spatial and temporal data, pre-processing and post-processing in data mining, robust and scalable statistical methods, security, privacy, and adversarial data mining, visual data mining, visual analytics, and data visualization.

Gang Liu · Eric Inae · Meng Jiang

# Deep Learning for Polymer Discovery

Foundation and Advances

Gang Liu
Department of Computer Science
and Engineering
University of Notre Dame
Notre Dame, IN, USA

Eric Inae
Department of Computer Science
and Engineering
University of Notre Dame
Notre Dame, IN, USA

Meng Jiang
Department of Computer Science
and Engineering
University of Notre Dame
Notre Dame, IN, USA

ISSN 2151-0067      ISSN 2151-0075 (electronic)
Synthesis Lectures on Data Mining and Knowledge Discovery
ISBN 978-3-031-84731-8      ISBN 978-3-031-84732-5 (eBook)
https://doi.org/10.1007/978-3-031-84732-5

© The Editor(s) (if applicable) and The Author(s), under exclusive license to Springer Nature Switzerland AG 2026

This work is subject to copyright. All rights are solely and exclusively licensed by the Publisher, whether the whole or part of the material is concerned, specifically the rights of translation, reprinting, reuse of illustrations, recitation, broadcasting, reproduction on microfilms or in any other physical way, and transmission or information storage and retrieval, electronic adaptation, computer software, or by similar or dissimilar methodology now known or hereafter developed.
The use of general descriptive names, registered names, trademarks, service marks, etc. in this publication does not imply, even in the absence of a specific statement, that such names are exempt from the relevant protective laws and regulations and therefore free for general use.
The publisher, the authors and the editors are safe to assume that the advice and information in this book are believed to be true and accurate at the date of publication. Neither the publisher nor the authors or the editors give a warranty, expressed or implied, with respect to the material contained herein or for any errors or omissions that may have been made. The publisher remains neutral with regard to jurisdictional claims in published maps and institutional affiliations.

This Springer imprint is published by the registered company Springer Nature Switzerland AG
The registered company address is: Gewerbestrasse 11, 6330 Cham, Switzerland

If disposing of this product, please recycle the paper.

# Preface

Modern polymeric materials have revolutionized various aspects of our lives, driving advancements in sustainability, biomaterials, biosensors, energy solutions, and aerospace technologies. The ever-evolving engineering and environmental challenges of our time demand materials with unconventional properties, such as high temperature stability, exceptional thermal conductivity, and biodegradability.

Polymers are composed of molecules represented as graph structures, where atoms act as nodes and bonds as edges. This inherent structure makes deep learning techniques—such as Transformers and Graph Neural Networks (GNNs)—indispensable for discovering new polymers that fulfill modern requirements.

Deep learning paradigms, namely prediction and generation, are integral to material virtual screening and inverse design, respectively. This book offers a systematic exploration of deep learning techniques tailored to polymer discovery, bridging the disciplines of materials science and artificial intelligence. It equips researchers and practitioners with foundational concepts and state-of-the-art methods for predicting polymer properties and designing novel polymers using advanced neural network architectures.

The content spans a broad spectrum of topics, progressing from fundamental concepts to advanced methodologies. It begins with polymer data representations and neural network architectures (Chap. 1) before delving into frameworks for property prediction (Chap. 2) and inverse polymer design (Chap. 3). Key approaches include sequence-based and graph-based techniques, leveraging neural network models such as LSTMs, GRUs, GCNs, and GINs. Advanced discussions encompass interpretable graph deep learning with environment-based augmentation (Chap. 4), semi-supervised methods for addressing label imbalance (Chap. 5), and data-centric transfer learning using generative methods like diffusion models (Chap. 6). Each topic is presented with detailed problem definitions, method descriptions, and experimental validations.

The book tackles pressing issues in polymer discovery, such as accurate property prediction, efficient design of polymers with desired traits, model interpretability, handling imbalanced and limited labeled data, and leveraging unlabeled data for enhanced predictions. Practical examples and experiments on real-world datasets demonstrate the efficacy of the proposed methodologies.

This book is designed for researchers, graduate students, and professionals in materials science, chemistry, and computer science who are interested in harnessing deep learning for polymer discovery and design. It serves as a primer, practical guide, and reference for those seeking to integrate artificial intelligence into materials research and development, inspiring innovation at the intersection of science and technology.

Notre Dame, USA  
December 2024

Gang Liu  
Eric Inae  
Meng Jiang

**Competing Interests** The authors declare the following as potential competing interests: This work was supported by NSF IIS-2142827, IIS-2146761, IIS-2234058, CBET-2332270, and ONR N00014-22-1-2507.

# Contents

**1 Polymer Data and Deep Neural Networks** .................... 1
  1.1 Polymer Data Representations .................... 1
    1.1.1 Sequences .................... 2
    1.1.2 Graphs .................... 2
    1.1.3 Vectors .................... 3
  1.2 Neural Networks .................... 4
    1.2.1 Modeling Tasks .................... 4
    1.2.2 Basic Neural Network Components .................... 6
    1.2.3 Advanced Neural Network Architecture .................... 8
  1.3 Foundation: Neural Network-Based Frameworks for Polymer Modeling .................... 10
    1.3.1 Deep Learning for Polymer Property Prediction .................... 10
    1.3.2 Deep Learning for Inverse Polymer Design .................... 12
  1.4 Advances: Deep Learning with Interpretable, Imbalance-Robust, and Generative Graph Methods .................... 13
    1.4.1 Interpretable Learning: Graph Rationalization with Environment-Based Augmentation .................... 13
    1.4.2 Imbalanced Learning: Semi-supervised Graph Imbalanced Regression .................... 14
    1.4.3 Generative Modeling: Data-Centric Learning from Unlabeled Graphs with Diffusion Model .................... 14
  References .................... 15

**2 Deep Learning for Polymer Property Prediction** .................... 17
  2.1 Problem Definition and Datasets .................... 17
    2.1.1 Polymer Property Classification .................... 17
    2.1.2 Polymer Property Regression .................... 18
    2.1.3 Dataset, Task Formulation and Evaluation .................... 18

- 2.2 Sequence(SMILES)-Based Prediction ... 20
  - 2.2.1 Recurrent Neural Networks ... 20
  - 2.2.2 LSTM ... 20
  - 2.2.3 GRU ... 21
  - 2.2.4 Transformer ... 22
- 2.3 Graph-Based Prediction ... 23
  - 2.3.1 Graph Neural Networks ... 23
  - 2.3.2 Graph Convolutional Networks (GCNs) ... 24
  - 2.3.3 GraphSAGE ... 25
  - 2.3.4 GAT ... 26
  - 2.3.5 GIN ... 26
  - 2.3.6 Graph Transformers ... 27
- 2.4 Specific Techniques ... 28
  - 2.4.1 Hyperparameter Tuning ... 28
  - 2.4.2 Data Augmentation ... 30
  - 2.4.3 Other Learning Paradigms ... 31
- 2.5 Summary ... 33
- References ... 34

# 3 Deep Learning for Inverse Polymer Design ... 37
- 3.1 Problem Definition and Datasets ... 37
  - 3.1.1 Unconditional Generation Without Constraints ... 37
  - 3.1.2 Conditional Generation With Constraints ... 38
  - 3.1.3 Datasets, Task Formulation and Evaluation ... 38
- 3.2 Generative Neural Network Architectures ... 39
  - 3.2.1 Generative Adversarial Networks ... 40
  - 3.2.2 Variational Autoencoders ... 40
  - 3.2.3 Diffusion Models ... 41
- 3.3 Unconstrained Polymer Generation ... 42
  - 3.3.1 Sequence-Based Generation ... 43
  - 3.3.2 Graph-Based Generation ... 45
- 3.4 Constrained Polymer Generation ... 47
  - 3.4.1 Constraint Types ... 47
  - 3.4.2 Adding Constraints as Features ... 50
- 3.5 Summary ... 51
- References ... 52

# 4 Interpretable Learning: Graph Rationalization with Environment-Based Augmentation ... 55
- 4.1 Introduction ... 55
- 4.2 Problem Definition ... 58
  - 4.2.1 Graph Property Prediction ... 58
  - 4.2.2 Graph Rationalization ... 59

|     | 4.3   | Interpretable Graph Neural Networks: GREA | 60 |
|---|---|---|---|
|     |       | 4.3.1 Rationale-Environment Separation | 60 |
|     |       | 4.3.2 Environment-Based Augmentations | 61 |
|     |       | 4.3.3 Optimization | 62 |
|     | 4.4   | Experiments | 63 |
|     |       | 4.4.1 Experimental Settings | 63 |
|     |       | 4.4.2 RQ1: Results on Effectiveness | 65 |
|     |       | 4.4.3 RQ2: Ablation Study on GREA | 65 |
|     |       | 4.4.4 RQ3: Case Study on Polymer Data | 65 |
|     |       | 4.4.5 RQ4: Results on Efficiency | 71 |
|     |       | 4.4.6 RQ5: Sensitivity Analysis | 72 |
|     | 4.5   | Conclusion | 72 |
|     |       | References | 74 |
| 5   | **Imbalanced Learning: Semi-Supervised Graph Imbalanced Regression** | | 77 |
|     | 5.1   | Introduction | 77 |
|     | 5.2   | Problem Definition | 79 |
|     | 5.3   | Self-Training Framework: SGIR | 80 |
|     |       | 5.3.1 A Self-Training Framework for Iteratively Balancing Scalar Label Data | 80 |
|     |       | 5.3.2 Balancing with Confidently Predicted Labels | 81 |
|     |       | 5.3.3 Balancing with Augmentation via Label-Anchored Mixup | 82 |
|     |       | 5.3.4 Optimization | 83 |
|     | 5.4   | Theoretical Motivations | 84 |
|     | 5.5   | Experiments | 88 |
|     |       | 5.5.1 Experimental Settings | 88 |
|     |       | 5.5.2 RQ1: Effectiveness on Property Prediction | 91 |
|     |       | 5.5.3 RQ2. Ablation Studies and Sensitivity Analysis | 96 |
|     | 5.6   | Conclusion | 98 |
|     |       | References | 100 |
| 6   | **Generative Modeling: Data-Centric Learning from Unlabeled Graphs with Diffusion Model** | | 103 |
|     | 6.1   | Introduction | 103 |
|     | 6.2   | Problem Definition | 106 |
|     | 6.3   | Data-Centric Transfer Framework: DCT | 107 |
|     |       | 6.3.1 Overview of Developed Framework | 107 |
|     |       | 6.3.2 Learning Data Distribution from Unlabeled Graphs | 107 |
|     |       | 6.3.3 Generating Task-Specific Labeled Graphs | 109 |

| | | | |
|---|---|---|---|
| 6.4 | Experiments | | 113 |
| | 6.4.1 | Experimental Settings | 113 |
| | 6.4.2 | RQ1: Outstanding Property Prediction Performance | 115 |
| | 6.4.3 | RQ2: Ablation Studies and Performance Analysis | 118 |
| | 6.4.4 | RQ3: Case Study for the Interpretability of Visible Knowledge Transfer | 120 |
| 6.5 | Conclusion | | 121 |
| References | | | 121 |

# Polymer Data and Deep Neural Networks

1

## 1.1 Polymer Data Representations

Polymers can be represented in multiple forms such as sequences, graphs, and vectors, as illustrated in Fig. 1.1. Neural networks have demonstrated their effectiveness in science by learning intricate relationships from data and leveraging this understanding to inform decision-making processes. When applying machine learning models (e.g., neural networks) to polymer tasks, the initial step involves determining the most suitable representation for the polymer data to be utilized by the models.

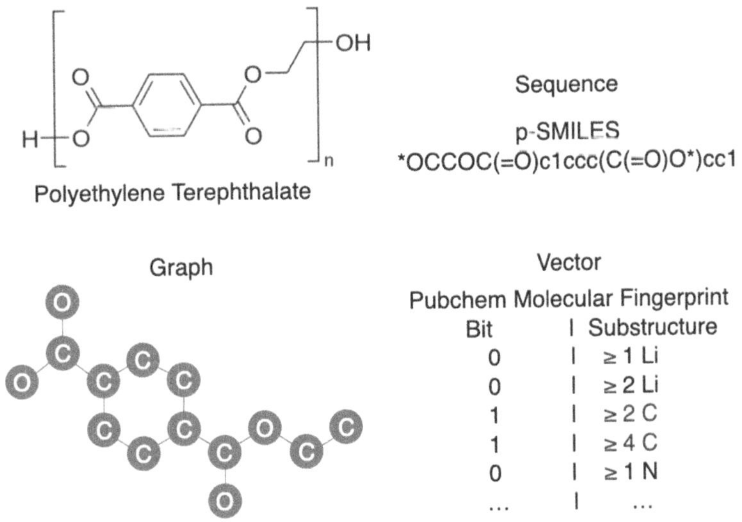

**Fig. 1.1** Visualization of a polymer's representations: **a** polyethylene terephthalate (PET); **b** sequence; **c** graph; **d** feature vector

© The Author(s), under exclusive license to Springer Nature Switzerland AG 2026
G. Liu et al., *Deep Learning for Polymer Discovery*, Synthesis Lectures on Data Mining and Knowledge Discovery, https://doi.org/10.1007/978-3-031-84732-5_1

## 1.1.1 Sequences

A sequence in machine learning is defined as data in which the points of the set are ordered and reliant upon the position of other points in the set. Formally, a sequence $S = [s_1, s_2, \ldots, s_n]$ is an ordered set of n points. For example, a molecular gene is a sequence of nucleotides in which the order is determined. In modeling molecules, there is not an obvious ordering to the atoms in the compound. However, several systems have been utilized to represent molecules and polymers as sequences. One such system is the simplified molecular-input line-entry system (SMILES), which is able to convert molecules to a sequence of atoms. Specific rules for conversion can be found in the following source. Because of the lack of an obvious atom ordering, there are typically many valid SMILES strings for a given molecule. Several algorithms have been proposed and used to canonicalize the choice of SMILES strings. Another system, called self-referencing embedded strings (SELFIES), is an alternate method for generating molecule sequences whose strength is that all SELFIES correspond to a valid molecule, whereas not all SMILES may be valid. SMILES can represent polymers, with the asterisk denoting the polymerization point, sometimes referred to as p-SMILES. BigSMILES is an extension of the standard SMILES representation which is able to better describe macromolecules. The main difference between SMILES and BigSMILES is the explicit representation of repeating structures. For example, *CC* and CC both represent the same segment of a carbon chain. However, the brackets in the BigSMILES representation CC define a repeating unit, while the asterisk is the p-SMILES representation *CC* shows a bonding site. This bonding site may be used to extend the carbon chain, or it may be used to bond to another compound. For most applications, SMILES (p-SMILES) is used for its popularity and will be the choice of sequence representation in later sections.

## 1.1.2 Graphs

A graph is a structure found within discrete mathematics that describes a set of objects and the relations between them. These objects in their most general form are called nodes or vertices, and the relationships between these objects are called edges or links. For example, social networks can be modeled as graphs by having users represent nodes and relationships between users represent edges. Similarly, molecules can be represented as graphs by letting atoms define nodes and chemical bonds define edges. Formally, we define a graph $G = (V, E)$ as a pair of sets $V$ and $E$, where $V$ is the set of nodes in the graph and $E$ is the set of node pairs, i.e., $\{v_1, v_2\}$ for $v_1, v_2 \in V$, which also represent edges.

## 1.1 Polymer Data Representations

Many graphs also contain additional node or edge information that further detail the graph. Returning to the social network graph, each user may have additional personal or demographic, such as name, age, and gender. Edges in the social network may contain details on the kind of relationship, such as the length or strength of the relationship. Molecular graphs similarly have different node and edge attributes to consider. Common node features include atomic number or chirality, and common edge features include bond type (e.g., single, double, or triple bond) or bond direction. As seen in Fig. 1.1c, we see the graph representation for polyethylene terephthalate. The graph still contains extra node and edge features such as double bonding, but this is usually maintained as a matrix of node and edge features.

### 1.1.3 Vectors

The standard definition for a vector is a quantity that has magnitude and direction. However, vectors in machine learning don't really contain a "direction". Rather, the vector is used to represent quantities in more than one dimension, similar to a vector defined using cartesian coordinates. Formally, a vector $a = [a_1, a_2, \ldots, a_n]$ is a sequence of values representing quantities in n dimensions. A simple feature vector of a molecule may be created by counting the occurrences of each type of atom in the molecule and then using each count to determine the magnitude of each feature dimension. For example, using polyethylene terephthalate, we may create the vector [2, 10, 5], where the three feature dimensions represent the occurrences of Hydrogen, Carbon, and Oxygen atoms. The choice of feature dimensions in the vector representation is often left to domain experts to decide by performing feature engineering. As previously mentioned, atom types and their occurrences may be used as features. However, this strategy loses important structural knowledge contained within the molecule, namely bond information. Another popular vectorization strategy for molecules is fingerprints, which broadly are defined as vectors created by exploring substructure fingerprints. Substructure fingerprints search the molecule for molecule fragments, such as Benzene rings or Hydroxyl groups, and record the number of occurrences of each fragment type. Substructure fingerprints have the advantage of better discriminating between molecules, but calculating each fingerprint may be slow, especially for large molecules, as graph matching is an NP-complete problem. The popular fingerprint methods are MACCS, ECFP, and Daylight Fingerprints (Mellor et al. 2019). Figure 1.1d represents the polymer polyethylene terephthalate as a vector using the PubChem Molecular Fingerprints (Kim et al. 2019).

## 1.2 Neural Networks

Neural networks, inspired by biological systems in the brain, are a framework for processing data and learning from them. In the mathematical sense, we can define neural networks as function $f$ that processes and updates itself using a data point $x$ and its label $y$.

### 1.2.1 Modeling Tasks

As depicted in Fig. 1.2, three tasks involve modeling the relationship between the data pair $(x, y)$. Classification and regression tasks fall under the broader prediction category. Here, the neural network processes the data point $x$ to produce its corresponding label. Conversely, the generation task differs from prediction tasks, aiming to generate a data point $x$. For all modeling tasks, the neural network is trained by comparing the output of the network to its expected output and backpropagating changes to the neurons to minimize these differences. The difference in expected and generated output can be quantified using a loss function $J$, where loss $L = J(y', y)$ for prediction and $L = J(x', x)$ for generation, where $y'$ is the predicted class.

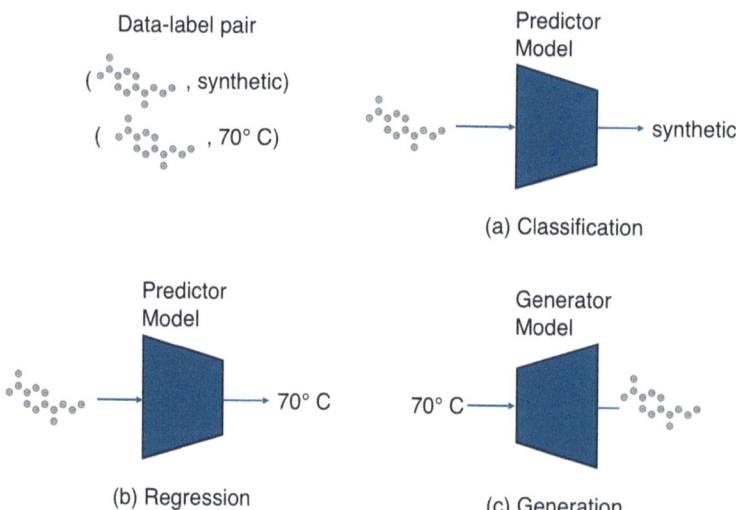

**Fig. 1.2** Visualization of different polymer modeling tasks: **a** classification on categorical values; **b** regression on numerical values; **c** conditional generation

## 1.2.1.1 Classification

The classification task can be modeled by a neural network $f : x \to y$, where the input to function $f$ is the data example $x$, and the target $y$ is a categorical value. In binary classification tasks, $y \in \{0, 1\}$ indicates that it belongs to a binary space with values of either 0 or 1. Cross-entropy loss, frequently applied to train neural networks, measures the discrepancy between the predicted probabilities (ranging from 0 to 1) and the actual labels. The closer the prediction probability is to the expected label, the less the cross-entropy loss. Formally, cross-entropy can be calculated by

$$L = -(y \log(p) + (1 - y) \log(1 - p)),$$

where $p$ is the predicted probability. For multi-class classification, the target space may contain more than two discrete label values. We could decompose the multi-class problem into multiple binary classification problems. Assuming the number of classes is $C$, the loss function becomes

$$\sum_{j=1}^{C} y_{i,j} \log(p_{i,j}),$$

where $y_{i,j}$ is a binary value determining if label $j$ is the correct class for the example $i$. $p_{i,j}$ is the predicted probability that the example $i$ is of class $j$. If a data example can belong to multiple classes, it is multi-class classification. In this case, the same cross-entropy loss function for multi-class classification can be used, with the only difference being $y_{i,j} = 1$ for two or more classes $j$.

## 1.2.1.2 Regression

Similar to classification, regression is also modeled by a neural network $f : x \to y$, label $y$ is a value within the real space of values $\mathbb{R}$. For regression, a common loss function is mean square error, or quadratic loss, which is defined as $L = (y' - y)^2$. Mean absolute error is another common loss function defined as $L = |y' - y|$. The strength of the mean square error is that large errors are more highly penalized. However, if the data is prone to outliers, then the mean absolute error may be more appropriate. In the case where $x$ has several regression labels, the network $f$ can be used to predict several regression values, and the previously described loss functions can be used on each prediction value.

## 1.2.1.3 Generation

The second modeling task discussed previously is generation, which can be described as a function $f : y \to x$, where the new generated sample should have $y$ as the associated label. The label $y$ can also be the empty set, which is known as unconditional generation. The neural network is usually trained by processing the data point $x$ into some latent representation

$z$, then recovering $x$. This trains the network to recognize important features during the reconstruction.

The loss function for the generation task can be defined along several metrics. For example, we can compare the reconstructed sample $x'$ with the original sample using mean squared error, $L = (x - x')^2$. We may also use the Kullback-Leibler (KL) divergence (Kingma 2013), which compares two data distributions. If the distribution of features in the original data diverges too much from the generated data, then the network will be penalized during training and updated to bring these two distributions into alignment.

### 1.2.2 Basic Neural Network Components

Neural networks process digital information by mimicking how biological networks process biological signals. The perceptron is a mathematical way to represent a biological neuron and serves as the most essential component for complex neural networks. The main idea of a perceptron component is to receive input information and then it either activates or not. Each network is composed of several perceptron components, organized into layers, and each layer is responsible for transforming input information through numerous steps, as depicted in Fig. 1.3, before passing the information to the next layer.

#### 1.2.2.1 Linear Layer

The first component of the perceptron is the linear layer. Linear layers can be described visually as two sets of neurons with neural connections between both sets. The input to each neuron is transformed using a linear equation before being passed to the output neuron. We can mathematically describe this behavior as a matrix transformation $\mathbf{w}$ on the data vector $\mathbf{x}$, where the output to the neurons is expressed through the equation $\mathbf{x}' = \mathbf{w}^T \mathbf{x}$.

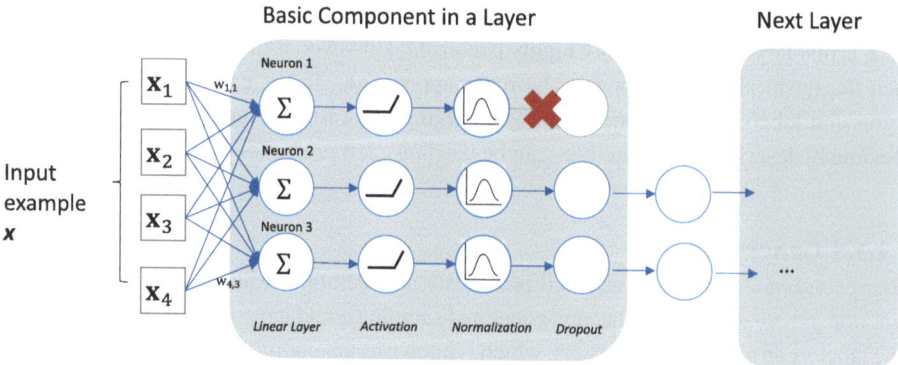

**Fig. 1.3** Visualization of basic neural network components

## 1.2 Neural Networks

In practice, linear layers may also include a bias term **b** to offset the data, leading to the full equation $\mathbf{x}' = \mathbf{w}^T\mathbf{x} + \mathbf{b}$. As dictated by linear algebra, compositions of linear transformations are also linear. To model non-linear relationships, we rely on the activation layer.

### 1.2.2.2 Activation Layer

The activation layer of a perceptron is a function that determines whether the neuron activates. Often this activation function is non-linear, as the non-linear activation between different linear layers is the key to the non-linear relationship modeling ability of neural networks.

An example activation function is a sigmoid function, defined as $\mathbf{x}' = (1 + e^{-\mathbf{x}})^{-1}$. Another function is the rectified linear unit (ReLU), defined as $\mathbf{x}' = \max(0, \mathbf{x})$. The ReLU is much simpler to compute than the sigmoid and converges more quickly when training.

### 1.2.2.3 Normalization Layer

After gathering the outputs from the activation layer, the neural network is left with a set of activations. However, the activation outputs may not be on the same scale as the inputs, potentially destabilizing neural network training (Ioffe 2015). If this is the case, then the next linear layer may need to adapt to deal with the larger or smaller magnitudes of the input. To eliminate this issue, a normalization layer is added to process and standardize the outputs of the previous layer. There are several types of normalization layers, all dependent on the variance of the inputs. Batch normalization subtracts the mean from the batch of inputs of the layer and divides it by the standard deviation. In this way, the inputs now have zero mean and unit variance. In practice, each layer does not necessarily require zero mean and unit variance, so batch normalization also adds two additional learnable parameters to the equation to generate the final output, which is used to scale and shift the output. The batch normalization can be described by the equation:

$$\mathbf{x}' = \gamma \left( \frac{\mathbf{x} - \mu}{\sqrt{\sigma^2 + \epsilon}} \right) + \beta,$$

where **x** is the input value, $\mu$ and $\sigma$ are the mean and standard deviation over the batch of input values, and $\gamma$ and $\beta$ are learnable parameters. $\epsilon$ is a small value for numerical stability.

To appropriately train from the batch, larger batch sizes are necessary. However, for inputs with high numerosity, the computation becomes intractable. Layer normalization alleviates the computational burden by calculating the mean and standard deviation over all features for each individual instance rather than over the whole batch. This also makes layer normalization independent of batch size, which makes layer normalization applicable to certain neural network architectures.

#### 1.2.2.4 Dropout Layer

When training the neural network, to encourage greater generalization and avoid overfitting, we may include a dropout layer (Krizhevsky et al. 2012). During training, individual outputs from the linear layer may be entirely dropped before passing on to the next layer. These outputs are set to zero according to a pre-defined probability.

### 1.2.3 Advanced Neural Network Architecture

A layer of neural network is not sufficient for complex real-world problems. Various neural network models are built upon basic perceptron components. The first layer of a neural network is typically called the input layer. It processes raw features into vector representations, which are then transformed and activated by subsequent layers of perceptrons. These layers and their intermediate output vectors are referred to as hidden layers and hidden representations, respectively, as their states are often hidden in a neural network model. The final output layer maps the final hidden representation (referred to as the latent representation) to the output space, where the loss function is calculated. Figure 1.4 provides an overview of various advanced neural network architectures designed to process sequences or graphs as input.

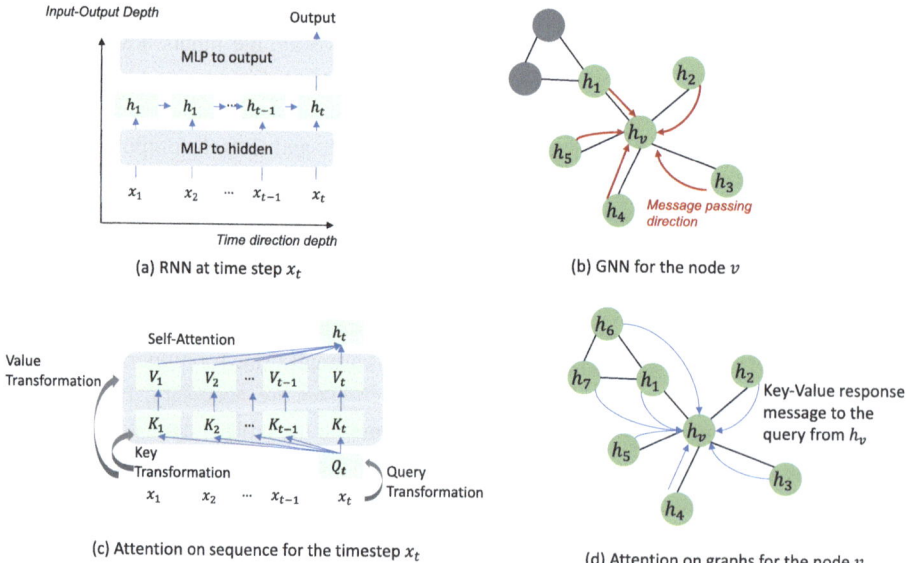

**Fig. 1.4** Visualization of four advanced neural network architectures. We can stack multiple layers of the basic component from Fig. 1.3 to construct MLP or FNN. We could utilize **a** RNNs for sequence data and **b** GNNs for graph data. Self-attention mechanism could be utilized for both **c** sequence data and **d** graph data

### 1.2.3.1 Feedforward Neural Networks

Feedforward artificial neural network (FNN) or multi-layer perceptron (MLP) models (Hornik et al. 1989) process each input data point individually with stacked layers of perceptrons. Besides the input and output layers, FNNs often include one or more hidden layers to capture non-linear relationships between input data and output labels. In practice, input relationships can be complex and reveal important recognition patterns. Therefore, we need more advanced architecture designs to capture these patterns.

### 1.2.3.2 Modeling Sequences with Recurrent Neural Networks

Sequence data, like time series and texts, comprise elements such as time steps or words. Neural networks that process sequential data must retain past information to make accurate future predictions. Recurrent neural networks (RNNs) are commonly employed to store this historical data within the sequence by updating hidden states (Hochreiter 1997).

At each time step $t$, RNNs maintain the hidden state of sequential memory $h_t$ of the current time step by incorporating information from the current input element $x_t$ and the previous hidden state, $h_{t-1}$, where $h_{t-1}$ maintains information from all previous information since the time $t-1$. So, $h_t$ extracts all historical information up to that time point.

This approach efficiently stores historical data in a hidden space and facilitates inference at different time points to avoid repeating calculations on historical data. The time direction depth of RNNs depends on the sequence length, while the input-output depth comes from stacking layers of hidden states $[h_t^1, h_t^2, \ldots, h_t^l]$, where $l$ is the index number of a layer. Each layer of the hidden state $h_t^l$ takes inputs from the previous layers, $h_t^{l-1}$, and the previous historical hidden state $h_{t-1}^l$.

In the next section, we will introduce in detail two variants of RNNs: Long Short-Term Memory Network (LSTM) and Gated Recurrent Unit (GRU), for processing sequence data.

### 1.2.3.3 Modeling Graph with Graph Neural Networks

Graph data comprises nodes and edges. To predict a node, a neural network must capture the linking relationships among its neighboring nodes. Graph Neural Networks (GNNs) aggregate information from local neighborhoods to update a node's hidden state (Kipf and Welling 2017). For a node $v$, a layer of GNN updates its hidden state $h_v$ by incorporating information from its neighbors $h_u$ based on the known edges in the graph. In a single layer of GNN, a node can only see information from its direct neighbors. Fortunately, GNN models are typically built with multiple layers, allowing nodes to iteratively aggregate information from neighbors. By deepening the GNNs, node $v$ can access broader neighbor information to update its own representation. However, deeper GNNs are notorious for suffering from an over-smoothing problem, which indicates that many aggregation iterations may make the hidden states of two connected nodes too similar. Thus, there is typically a balance between the depth of the GNN and the visible range of a node. In the next section, we will introduce three GNN variants: Graph Convolutional Networks (GCNs), Graph Attention Networks (GAT), and Graph Isomorphism Networks (GIN).

### 1.2.3.4 Attention Mechanism

The attention mechanism is the core idea behind the Transformer, which lays the groundwork for recent powerful chat models like GPT-4. It relies on three variables—queries, keys, and values—to search for important keys matching the query variables and then retrieve corresponding values based on the attention results. For instance, given a query element 'nice' in a sequence context like "Today is nice," and three key elements ["time", "fruit", "name"] with three corresponding value elements ["day", "apple", "Bob"], the query element may match different keys resulting in attention scores [0.7, 0.1, 0.2]. These scores help identify the correct value element, such as "day," with the highest probability. The most common approach to obtaining query, key, and value variables is through self-attention, where three linear transformations are used to obtain three variables on the same data. For sequential data, these transformations are applied to elements in each sequential timestep, with each timestep having corresponding query, key, and value components. In graph data, transformations are applied to all nodes. In the following section, we will introduce how to apply the attention mechanism, especially self-attention and Transformer for sequence and graph learning, as well as polymer discovery.

## 1.3 Foundation: Neural Network-Based Frameworks for Polymer Modeling

Discovering polymers involves analyzing the relationships between polymer structure and properties. Structural features could be denoted as variable $x$, while property values are denoted as variable $y$. Machine learning frameworks with neural network models can be developed based on assumptions regarding $x$, its availability, and its relationship with $y$. For predicting property values when $x$ is available, a polymer property prediction framework is utilized. Conversely, when generating structure $x$ from noisy data with conditional label $y$, it is referred to as inverse polymer design.

### 1.3.1 Deep Learning for Polymer Property Prediction

Predicting polymer properties necessitates the transformation of polymer features and structures into hidden vectors, enabling neural networks to effectively process them. These networks encode the inputs into a polymer's latent representation, often represented as a matrix. This condensed representation of the polymer sequence or graph is then transformed into

## 1.3 Foundation: Neural Network-Based Frameworks for Polymer Modeling

a single vector for output prediction. Finally, the optimization of neural networks is guided by a loss function. We present an overview below, with more details described in Chap. 2.

#### 1.3.1.1 Input Processing

Polymers can be represented in various forms such as SMILES sequences, molecular graph structures, or fingerprint vectors, each requiring the framework to process them differently.

The input sequence is typically a SMILES string like "FCC1CC2CC1C(*)C2*", with the asterisk ("*") representing polymerization points. The initial step involves tokenizing the symbol sequence by breaking it into minimal units for neural network processing. These minimal units are also referred to as tokens. This can be done based on the chemical elements, resulting in a sequence of unit symbols ['F', 'C', 'C', ..., '*']. Next, the embedding vector for each unit is indexed from an embedding table. This table is randomly initialized and then optimized.

The input graph of molecules has two types of units to be processed by the neural network: nodes (atoms) and edges (bonds). Each node has associated features like atom type, chirality, and formal charge, while each bond has features like bond type and bond stereo. An atom encoder converts atom features into initial node embeddings, and a bond encoder converts bond features into bond embeddings. These initialized embeddings are then used in subsequent layers of graph neural networks.

Processing polymer vector input is simpler compared to sequence and graph inputs. It resembles the input layer of MLP models with a linear layer to convert the input vector into an initial embedding for further processing.

Besides, feature normalization and standardization are common techniques used to improve initialized embeddings. These techniques generally stabilize the model training process by regularizing the initial embeddings.

#### 1.3.1.2 Feature Encoding

Multiple layers of neural networks can be stacked to capture complex relationships between input features and target properties. After processing the input, RNNs, GNNs, or MLPs can encode SMILES strings, polymer graphs, or fingerprint vectors into latent spaces, respectively. RNNs maintain the hidden state from the first token in the sequence to the last one (Nazarova et al. 2021).

There are numerous hyperparameters to tune when searching for the best neural network model architecture, such as the number of hidden states (i.e., the dimension of hidden representation) and the depth of the layers. Additionally, there are several simple and effective techniques to stabilize the training process and deepen neural network models, including batch normalization, layer normalization, dropouts, and residual connections. Details will be introduced in the next section.

### 1.3.1.3 Representation Summarization

RNNs and GNNs learn the representation on more fine-grained units (i.e., atom level). Because the property is associated with the polymer, it is necessary to summarize the essential representations for the entire polymer structure.

In RNNs, the summarization operation is often associated with a special token. A SMILES string, similar to a sentence in natural language, can have a special token called [CLS] inserted at the beginning. The encoding vector of [CLS] is commonly used as the "sentence vector," serving as the summarization vector for the entire SMILES string for property prediction.

In GNNs, the summarization operation, also known as the Readout function, condenses a matrix of node representations into a vector of polymer representation. The Readout function can be defined in various ways, ranging from simple averaging and summing to more complex adaptive and learning-based approaches (Liu et al. 2024).

### 1.3.1.4 Output and Optimization

After encoding and summarizing, the output layer converts the latent representation of the neural network to the final output, either categorical or numerical values. During inference/testing, the outcome is the neural network prediction result. During training, the loss function computes the error between the output and the target based on the types of modeling tasks.

## 1.3.2 Deep Learning for Inverse Polymer Design

The inverse design task involves generating polymers, possibly based on property or structure constraints. Two key differences between inverse design and property prediction are input processing and output decoding. Neural networks process input information and are trained to generate desirable polymer structures as output. We present an overview below, with more details described in Chap. 3.

### 1.3.2.1 Input Processing

Inputs serve as constraints for the neural network generation process. When input data is random noise from a known distribution, we refer to this as unconditional or unconstrained generation. Specifying constraints such as polymer substructures, categorical or numerical numbers, or their combinations is desirable as it makes the generation process more controllable. A neural network module could handle these constraints by mapping them to vectors. For instance, RNNs and GNNs can process sequence-related or graph-related polymer substructures, respectively, while MLPs can encode categorical and numerical values.

#### 1.3.2.2 Feature Decoding and Optimization

Feature encoding in property prediction extracts a concise latent representation from the input, while generation tasks decode fine-grained information for the final polymer structures based on input constraints. Deepening the hidden layers of RNNs, GNNs, or MLPs is preferable in the decoding module to capture increasingly fine-grained information with the network's depth. Similar to the feature encoding architecture, the choice of a specific neural network architecture depends on the assumption of the polymer structure, such as SMILES-style sequences or graphs. The reconstruction loss defined in the previous generation task section is often used to optimize the decoding module with high-quality decoding information.

#### 1.3.2.3 Post-generation Processing

The generation task typically involves post-processing procedures to transform the output from the feature decoding layer into token unit identifiers. These identifiers are then processed to complete SMILES or polymer graphs.

For sequence data, the neural network decodes features often as indexes of tokens in the embedding table. The post-processing should map these indexes to their real meanings, such as carbon or nitrogen elements.

For graph data, the decoded features include both node and edge features (i.e., the adjacency matrix indicating connected nodes and edge types). The post-processing should convert nodes and edges back to real atom and bond types and recover the connection information between atoms.

## 1.4 Advances: Deep Learning with Interpretable, Imbalance-Robust, and Generative Graph Methods

Deep learning methods for polymer discovery often lack transparency and interpretability in decision-making and suffer from limited supervision due to the time-consuming nature of material data collection. It often takes months or years to synthesize and annotate only a few examples, resulting in scarce labeled data for training supervised GNN models. To address these issues,

### 1.4.1 Interpretable Learning: Graph Rationalization with Environment-Based Augmentation

Rationale is defined as a subset of input features that best explains or supports the prediction by machine learning models. Rationale identification has improved the generalizability and interpretability of neural networks on vision and language data. In graph applications such as molecule and polymer property prediction, identifying representative subgraph structures named graph rationales plays an essential role in the performance of graph neural networks. Existing graph pooling and/or distribution intervention methods suffer from lack of examples to learn to identify optimal graph rationales. To advance interpretable learning in polymer discovery, in Chap. 4, we present a data augmentation-based method to improve graph rationale methods for polymers.

We will introduce an augmentation operation called *environment replacement* that automatically creates virtual data examples to improve rationale identification. There is an efficient framework that performs rationale-environment separation and representation learning on the real and augmented examples in *latent spaces* to avoid the high complexity of explicit graph decoding and encoding. Comparing against other techniques, experiments on seven molecular and four polymer real datasets demonstrate the effectiveness and efficiency of the augmentation-based graph rationalization framework.

### 1.4.2 Imbalanced Learning: Semi-supervised Graph Imbalanced Regression

Data imbalance is easily found in annotated data when the observations of certain continuous label values are difficult to collect for regression tasks. When they come to molecule and polymer property predictions, the annotated graph datasets are often small because labeling them requires expensive equipment and effort. To address the lack of examples of rare label values in graph regression tasks, in Chap. 5, we will introduce a semi-supervised framework to progressively balance training data and reduce model bias via self-training.

The training data balance is achieved by (1) pseudo-labeling more graphs for under-represented labels with a novel regression confidence measurement and (2) augmenting graph examples in latent space for remaining rare labels after data balancing with pseudo-labels. The former is to identify quality examples from unlabeled data whose labels are confidently predicted and sample a subset of them with a reverse distribution from the imbalanced annotated data. The latter collaborates with the former to target a perfect balance using a novel label-anchored mixup algorithm. We show experiment results in seven regression tasks. These results demonstrate that the imbalanced learning framework significantly reduces the error of predicted molecular and polymer properties, especially in under-represented label areas.

### 1.4.3 Generative Modeling: Data-Centric Learning from Unlabeled Graphs with Diffusion Model

Graph property prediction tasks are important and numerous. While each task offers a small size of labeled examples, unlabeled graphs have been collected from various sources and at a large scale. A conventional approach is training a model with the unlabeled graphs on self-supervised tasks and then fine-tuning the model on the prediction tasks. However, the self-supervised task knowledge could not be aligned or sometimes conflicted with what the predictions needed. In Chap. 6, we will extract the knowledge underlying the large set of unlabeled graphs as a specific set of useful data points to augment each property prediction model.

We will use a diffusion model to fully utilize the unlabeled graphs and two objectives to guide the model's denoising process with each task's labeled data to generate task-specific graph examples and their labels. Experiments demonstrate that the data-centric approach performs significantly better than fifteen existing various methods on fifteen tasks. The performance improvement brought by unlabeled data is *visible* as the generated labeled examples unlike the self-supervised learning.

## References

S. Hochreiter. Long short-term memory. *Neural Computation MIT-Press*, 1997.

K. Hornik, M. Stinchcombe, and H. White. Multilayer feedforward networks are universal approximators. *Neural networks*, 2(5):359–366, 1989.

S. Ioffe. Batch normalization: Accelerating deep network training by reducing internal covariate shift. *arXiv preprint* arXiv:1502.03167, 2015.

S. Kim, J. Chen, T. Cheng, A. Gindulyte, J. He, S. He, Q. Li, B. A. Shoemaker, P. A. Thiessen, B. Yu, et al. Pubchem 2019 update: improved access to chemical data. *Nucleic acids research*, 47(D1):D1102–D1109, 2019.

D. P. Kingma. Auto-encoding variational bayes. *arXiv preprint* arXiv:1312.6114, 2013.

T. N. Kipf and M. Welling. Semi-supervised classification with graph convolutional networks. In *International Conference on Learning Representations*, 2017.

A. Krizhevsky, I. Sutskever, and G. E. Hinton. Imagenet classification with deep convolutional neural networks. *Advances in neural information processing systems*, 25, 2012.

G. Liu, E. Inae, T. Luo, and M. Jiang. Rationalizing graph neural networks with data augmentation. *ACM Transactions on Knowledge Discovery from Data*, 18(4):1–23, 2024.

C. Mellor, R. M. Robinson, R. Benigni, D. Ebbrell, S. Enoch, J. Firman, J. Madden, G. Pawar, C. Yang, and M. Cronin. Molecular fingerprint-derived similarity measures for toxicological read-across: Recommendations for optimal use. *Regulatory Toxicology and Pharmacology*, 101:121–134, 2019.

A. L. Nazarova, L. Yang, K. Liu, A. Mishra, R. K. Kalia, K.-i. Nomura, A. Nakano, P. Vashishta, and P. Rajak. Dielectric polymer property prediction using recurrent neural networks with optimizations. *Journal of Chemical Information and Modeling*, 61(5):2175–2186, 2021.

# Deep Learning for Polymer Property Prediction

## 2.1 Problem Definition and Datasets

In the previous chapter, we have discussed various ways to model polymers using data representations and how neural networks can be utilized to process the data. The neural network modeling of polymers can be used to predict the properties of polymers. There are two types of prediction tasks we can consider: classification, in which polymers are grouped into discrete classes (i.e. toxic or non-toxic), and regression, in which polymer properties are given as continuous values (i.e. melting point). Given training data points, consisting of a polymer and its label value, one can train a neural network to predict the properties of new polymers. In this chapter, we will discuss the problem of polymer property prediction in addition to several popular neural network architectures for both classification and regression prediction tasks.

### 2.1.1 Polymer Property Classification

Some polymer properties are categorical, making the polymers belong to two or more distinct classes. Polymer classification refers to using neural networks to predict discrete property values. There are two types of classification problems depending on the number of possible label values. Tasks with only two class labels are called binary classification, whereas tasks with more than two classes are called multi-class classification. We can further distinguish tasks where a single polymer may have multiple valid class labels as multi-label classification, though these tasks are less common. Examples of polymer classification tasks include solvent compatibility (Chandrasekaran et al. 2020) and toxicity (Kumar et al. 2020). It is important to note that the choice of prediction task is often dependent on the availability of training data. Due to the expenses associated with obtaining label values through chemical experimentation, training datasets may only contain several thousand available samples, despite the polymer design space being incredibly vast (Kuenneth and Ramprasad 2023).

### 2.1.2 Polymer Property Regression

The other type of polymer property is those that are continuous. These values are defined on the real value space, meaning that the set of possible labels is technically infinite and not confined to a finite discrete set. For example, the glass transition temperature for polymers can potentially be any real number $\mathbb{R}$, with typical values between $-100$ to $200\,°C$ (Otsuka et al. 2011).

There exist many other properties that can be measured and predicted, including density, heat of fusion, thermal conductivity, and many other physical, permeability, or solubility measurements, due to the large number of applications for polymers (Otsuka et al. 2011). As mentioned previously, the choice of regression task is also constrained due to polymer data scarcity.

### 2.1.3 Dataset, Task Formulation and Evaluation

The datasets for property prediction are composed of data/label pairs, where polymer structures are given with their associated property value. Specific polymer data repositories include PoLyInfo (Otsuka et al. 2011), MSA (Thornton et al. 2012), and the CHiMaD (Project 2025) database. Each polymer example can be represented using p-SMILES, which is a SMILES format that represents polymers as individual monomers with indicators to designate polymerization points. For example, the compound poly(ethylene terephthalate) is represented as "*OCCOC(=O)c1ccc(C(=O)O*)cc1," where the asterisks denote polymerization points.

For property prediction, the polymer data representation is used as input to a prediction model to generate some output label value. Formally, we say that given a polymer $x$ and its associated label property $y$, the goal is to train a neural network $f(x)$ that predicts a value $y'$. Because the goal of property prediction is to correctly predict the label value $y$, we train the neural network to reduce the error between $y$ and its prediction $y'$. For classification, this means reducing the probability of predicting an incorrect class. For regression, this means reducing the numerical distance between the two values.

The way to evaluate the performance of the neural network is based on the prediction task, classification or regression. For binary classification, in which there are only two label classes, the most common evaluation metric is ROC-AUC, which stands for Receiver Operating Characteristic—Area Under the Curve. The ROC is a probability curve that plots the TPR (true positive rate) of the model against the FPR (false positive rate) at various threshold settings. The AUC represents the area under the ROC curve, which ranges between

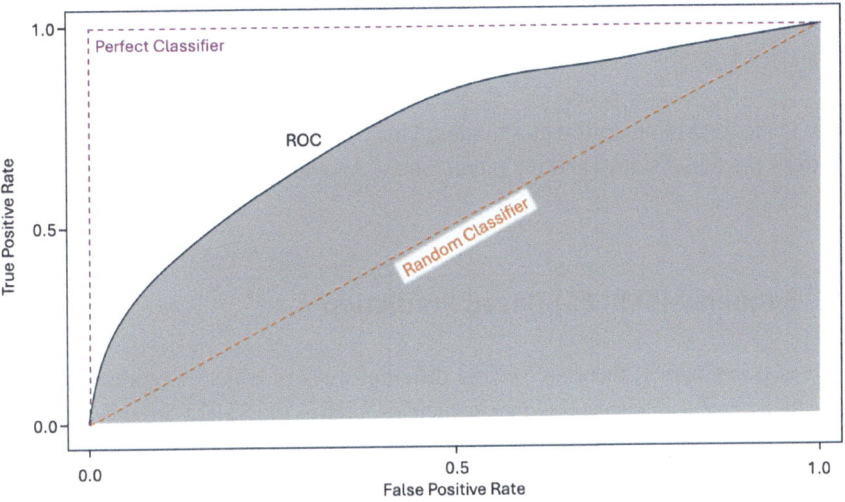

**Fig. 2.1** An example of ROC-AUC curves

0 to 1, with higher values being associated with better model performance. If the ROC-AUC is 1, this implies that the TPR is always 1 and the model correctly predicts classification labels at all thresholds. This is modeled in Fig. 2.1 with the "perfect classifier" curve. An ROC-AUC of 0.5 implies that the TPR and FPR are about the same at each threshold and the model has a 50/50 chance of correctly predicting any of the properties, which is modeled by the "random classifier" in the figure. For classification tasks with greater than two classes, it is still possible to use ROC curves using a One-vs-All evaluation, where the false positive rate encompasses all incorrect class predictions. A more common multi-class evaluation is Cross-Entropy, where we measure the difference between the probability distribution of a given prediction and the actual target label. Essentially, the more confidently the model can generate a correct classification, the better the evaluation. Lower Cross-Entropy scores are preferred for this evaluation.

For regression, one common metric is MAE (mean absolute error), in which the average distance between each model prediction $y'$ and label value $y$ is calculated. We can represent MAE using the equation:

$$\text{MAE} = \frac{1}{n} \sum_{i=1}^{n} |y_i - y'_i|,$$

where $n$ is the total number of data samples. A lower MAE demonstrates that the model's predicted values closely match the actual labels.

Another common evaluation metric is RMSE (root mean squared error), which is useful as it more heavily penalizes larger prediction discrepancies. RMSE can be calculated using the following equation:

$$\text{RMSE} = \sqrt{\frac{1}{n} \sum_{i=1}^{n} (y_i - y_i')^2},$$

where $n$ is the total number of data samples. This metric measures the difference between the model's predicted values and the target labels. As with MAE, lower RMSE values are desired.

## 2.2 Sequence(SMILES)-Based Prediction

As discussed in Chap. 1, there are several different ways in which to represent polymers, one of which being sequences. Polymer sequences, such as the SMILES representation, can be fed as input to neural networks to generate output prediction. In this section, we discuss two methods for sequence-based polymer predictions: Recurrent Neural Networks (RNNs) and Transformers.

### 2.2.1 Recurrent Neural Networks

Recurrent Neural Networks (RNNs) are a type of neural network built to process sequential data. As opposed to traditional neural networks, in which inputs and outputs are assumed to be independent of each other, the RNN takes into account previous inputs as they relate to the current input. This "memory" makes RNNs applicable to sequential data, where the sequence ordering is vital for prediction. For example, the order of atoms within a SMILES representation of a polymer conveys necessary atomic bonding information. Should each atom be processed without knowledge of bonds, then structural isomers, which contain the same number of atoms of each element but have distinct bonding configurations, will be virtually identical.

The structure of an RNN, as seen in Fig. 2.2a, resembles a chain of units, in which the inputs from each layer are processed and passed on to the next layer. Each recurrent unit, Fig. 2.2b, in addition to passing output to the next layer, also passes the output to the next unit in the recurrent unit chain. Through this process, units in the chain will receive information from previous units in the chain, allowing for previous inputs to inform latter ones. After the input has been processed using several RNN layers, we apply a final output layer to generate a prediction for the polymer. This output layer may be a multi-layer perceptron (MLP) that has prediction values (or class prediction probabilities) as explicit outputs.

### 2.2.2 LSTM

A traditional RNN, while sufficient for short sequences, is typically insufficient to remember previous inputs over long periods. Long Short-Term Memory (LSTMs), which are a type of RNN capable of remembering these long-term dependencies, addresses this limitation

## 2.2 Sequence(SMILES)-Based Prediction

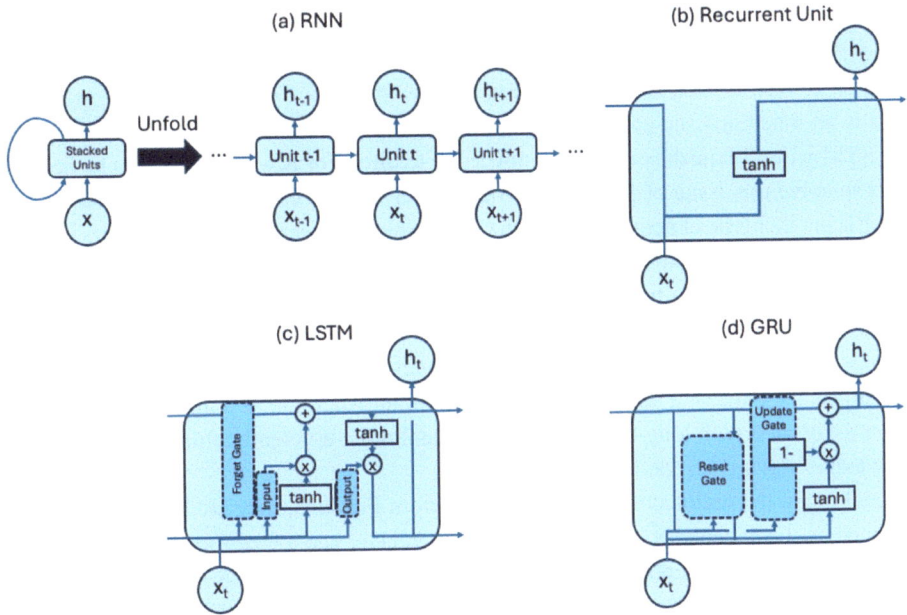

**Fig. 2.2** Visualization of Recurrent Neural Network Architectures. RNNs are constructed as a chain of repeating modules (**a**). These modules can be a standard recurrent unit (**b**), or contain more detailed structures such as the long short-term memory unit (**c**) or gated recurrent unit (**d**)

of traditional RNNs. The core difference of LSTMs as opposed to traditional RNNs is the inclusion of a structure called a gate. As seen in Fig. 2.2c, the LSTM unit contains a path that appears to flow through all units with minimal modifications. This path essentially chooses what information is allowed to be "remembered" and passed through the unit chain using an input gate and what information should be "forgotten" using a forget gate. In this way, information can be passed from previous units virtually unchanged without the risk of losing information via a traditional RNN unit chain. A third output gate is used to determine the output of the LSTM unit to be sent to the next layer of the RNN.

### 2.2.3 GRU

Gated Recurrent Units (GRUs) are another type of RNN which also include a gating mechanism to pass information between units, which we show in Fig. 2.2d. The GRU contains an update gate and reset gate, similar to the LSTM's input and forget gates. However, the GRU differs from the LSTM as it does not contain an additional output gate. This difference limits the GRU from being capable of understanding some complex data patterns, but it also reduces the number of parameters that need to be trained.

### 2.2.4 Transformer

However powerful LSTMs and GRUs are in allowing for a form of "memory" in RNNs, there is an inherent issue called catastrophic forgetting. At some point, the forget gate of an LSTM will relinquish some prior knowledge to allow for new knowledge to be held. In order to solve this issue of RNNs, the Transformer was developed (Vaswani et al. 2017), which is another type of neural network that is capable of processing sequential data without the same information bottleneck problem. Powerful large language models (LLMs) such as ChatGPT (generative pre-trained transformer) (Achiam et al. 2023) and BERT (Bidirectional encoder representations from transformers) (Devlin et al. 2018) utilize transformers heavily to process sequential text data. The main mechanism behind transformers is attention, which determines the parts of a sequence that are important. Attention as a mechanism mitigates issues associated with long-term dependencies and more efficiently utilizes this long-term knowledge, solving several issues of RNNs.

Looking at the architecture of transformers from Fig. 2.3a, they are built using a stack of encoders. Each encoder contains a self-attention layer and a feed forward network, such as an MLP. The self-attention layer determines which elements of the input are most associated with each other. For example, the p-SMILES sequence for poly(ethylene terephthalate), "OCCOC(=O)c1ccc(C=O)O*)cc1", contains a benzene ring which is represented by "c1ccccc1" in the string. While the two "c1"s are separated in the sequential representation, these two carbon atoms are, in reality, connected via a single bond. The self-attention layer would be able to recognize this association and encode each atom into a new representation

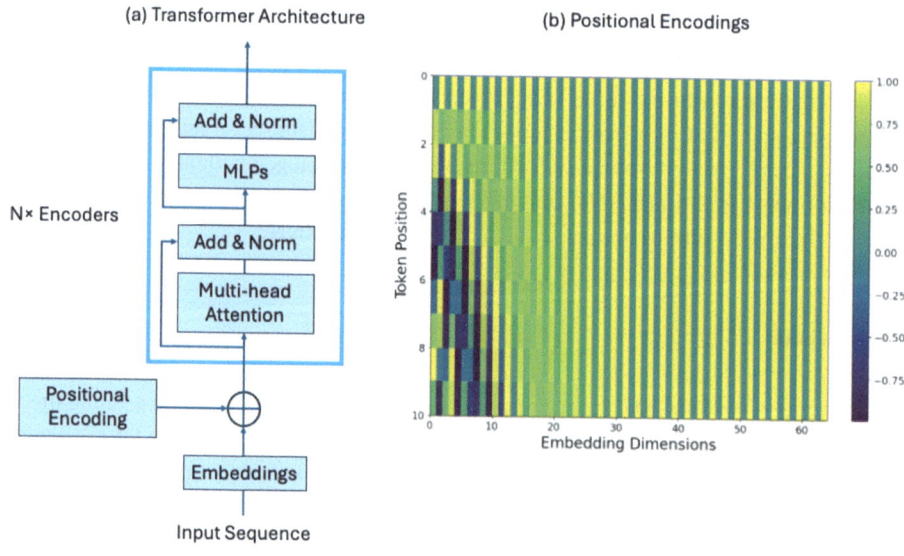

**Fig. 2.3** Visualization of a transformer architecture (**a**) and positional encodings (**b**)

with this knowledge. It is reminiscent of the memory mechanism of RNNs and its variants but is able to directly access relevant information instead of having it pass through a series of gates. Additionally, transformers process the entire sequence at once. This allows transformers to be much quicker than LSTMs, but this means that the positional knowledge of the input sequence is lost. To solve this problem, transformers will add positional information to the embeddings to provide the self-attention layer with spatial knowledge of the input sequence, improving the encoder's ability to recognize common patterns that may appear in the sequential data. These positional encodings are called sinusoidal PEs, and they are based on a series of sine waves. Figure 2.3b contains a visualization of sinusoidal PEs, and Returning to the benzene ring example, because all benzene rings contain six atoms, positional encodings may help the self-attention layer recognize the beginning and end of the ring as they must be separated by four other carbon atoms. After both the self-attention and feed forward layers, the encoder adds the output of the layer to the input via a residual connection and performs normalization. For typical transformer models, including polymer generation models, the transformer will include a stack of decoders as well. These decoders can generate new text and polymers depending on the application. However, for use in discriminative models, such as polymer classification and regression tasks, only the encoders are strictly necessary. SMILES-BERT (Wang et al. 2019), a transformer-based prediction model for polymer sequences, only uses the encoder stack from its base model BERT.

## 2.3 Graph-Based Prediction

While sequence-based polymer prediction methods have seen steady development as described in the previous sections, it is more common to represent polymers as graphs, as graphs are a more natural representation of molecules. In fact, chemical graph theory has a long history dating back to the 1800s with mathematicians Arthur Cayley and James Sylvester (Rouvray 1989). In this section, we introduce graph based strategies for polymer prediction, including several types of graph neural networks (GNNs) and Graph Transformers.

### 2.3.1 Graph Neural Networks

Graph Neural Networks (GNNs) are a type of neural network specifically designed to leverage the graph structure in the learning process. The main feature of GNNs is message passing, in which graph nodes update their representations based on the aggregation of messages from their neighbors. For this reason, GNNs may also be referred to as message passing neural networks (MGNN).

To perform message passing, the GNN will first embed the node's features (such as element or chirality) into a latent representation. Then, the node's initial representation will be passed to its direct neighbors and aggregated. There are several choices for aggregation strategies, several of which will be introduced in the following sections.

The general aggregation and update scheme of GNNs can be described using these two equations:

$$\mathbf{h}_{N(i)} = \text{aggregate}(\{\mathbf{x}_j : j \in N(i)\})$$

$$\mathbf{h}_i = \text{combine}(\{\mathbf{x}_i, \mathbf{h}_{N(i)}\}),$$

where $\mathbf{x}_i$ and $\mathbf{x}_j$ are node features for node $i$ and node $j$, $j$ is taken from the neighborhood $N(i)$ of $i$, and $\mathbf{h}_i$ is the node representation of $i$ after one GNN layer. The GNN may have multiple layers, which means that the node's new updated representation is then passed to its neighbors once again and aggregated to create new embeddings. Once the inputs have been processed by all GNN layers, we then aggregate all of the node representations using a global pooling strategy to generate a single graph representation. Graph pooling is most typically performed by either finding the element-wise mean or sum of the latent values across all node representations.

One important feature to note about GNNs is its permutation invariance. Invariance refers to a function whose output is the same regardless of the ordering of the input. Formally, this means that a function f is invariant if $f(S) = f(\pi_S(S))$ for all permutation functions $\pi_S$ on the input set $S$. This means that a GNN will generate the same graph representation no matter the order in which the nodes are processed. This strength of GNNs solves the issue of sequence-based polymer representations, in which not all SMILES representations of the same polymer may be treated equally by an RNN or Transformer. There are several different architectures for GNNs which we will discuss in this section, the difference between them being their choice in aggregation and combination functions.

## 2.3.2 Graph Convolutional Networks (GCNs)

Graph Convolutional Networks (GCNs) are a type of graph neural network inspired by Convolutional Neural Networks (CNNs), typically used to process image data (Kipf and Welling 2017). In a CNN, a pixel and its neighbors are processed together, such as taking the sum of each pixel value, to create representations for small visual features. A GCN works similarly to CNNs except that the number of neighbors processed by each neural network is not fixed. A pixel in an image will have exactly eight neighbors (aside from border pixels), whereas a node in a graph may have any number of node neighbors.

To solve this problem, we can divide the sum of neighbor features by the degree of each node, finding the mean of the neighbors' features. Another consideration is that nodes with higher degrees are able to propagate their features more easily than nodes with smaller degrees. Therefore, we can add a weight term that reduces the influence of nodes with higher degrees.

## 2.3 Graph-Based Prediction

We can model one layer of graph convolution using the equation:

$$h_i = \sum_{j \in N_i} \frac{1}{\sqrt{\deg(i)\deg(j)}} W x_j,$$

where features $x_j$ of node $j$, taken from the neighborhood $N_i$ of node $i$, are multiplied by a learnable matrix $W$ and averaged according to the degrees of $i$ and $j$ to create node representations $h_i$.

In essence, one layer of graph convolution is a linear layer of a standard neural network but adjusted to consider node neighborhoods. As mentioned previously, GNNs can create a final graph representation using graph pooling, so the node representations $h_i$ will be element-wise averaged or summed together to create a final graph representation $h_G$ for graph $G$.

### 2.3.3 GraphSAGE

The traditional data flow of GCNs requires the use of an adjacency matrix as input, which is a graph representation designating node connections. However, the adjacency matrix is fixed as input, the GCN is unable to adapt to unseen nodes that don't exist in the training graph. Furthermore, convolutional layers are ill-suited for large graphs with a high quantity of nodes. The scalability of GNNs to large graphs was then addressed by GraphSAGE (Hamilton et al. 2017), another type of graph neural network that is typically used for large graphs in inductive settings, allowing for generalization to previously unseen graphs.

The difference that facilitates learning on large graphs is neighborhood sampling, which occurs before feature aggregation, hence the name (Graph SAmple and aggreGatE). The neighborhood $N(i)$ for node $i$, rather than being every single neighbor of $i$, is a fixed-size set of neighbors.

Another difference is that, before updating the node's representation, GraphSAGE concatenates the node representation with the values of the aggregated neighborhood for the target node. Note that the aggregation function for GraphSAGE is the maximum function, rather than the mean.

We can thereby formalize the GraphSAGE function as the equation:

$$h_i = W \cdot \text{concat}(x_i, \max(\{x_j : \forall j \in N(i)\})),$$

where $x_i$ is the node features for node $i$, $x_j$ are node features for all nodes $j$ in the sampled neighborhood $N(i)$, $W$ is a learnable matrix, and concat($\cdot$) and max($\cdot$) are the concatenation and maximum functions, respectively.

### 2.3.4 GAT

With the development of the attention mechanism, which had been previously developed for text, scientists wondered if attention could similarly be used for graphs. GCNs and GraphSAGE use fixed algorithms, such as calculating node degree, to determine node importance during aggregation. However, it intuitively makes sense that node importance cannot be fully represented using fixed aggregation strategies. This conjecture led to the development of Graph Attention Networks (GATs), which are a type of network that leverages the idea of attention from Transformers to process graph-structured data. GATs are built using a similar framework as GCNs, except that the weight term used to find the weighted average is not predefined. To appropriately determine the importance of each neighbor node to its target during each layer of convolution, the GAT will leverage a single self-attention layer across node features for nodes within the neighborhood. In practice, this layer can be a single FNN layer, such as an MLP.

The weight determined by self-attention is then normalized using the softmax function. The equation for generating node representations using a GAT layer can then be defined as:

$$\mathbf{h}_i = \sum_{j \in N(i)} \frac{\exp(\text{MLP}(\mathbf{x}_i, \mathbf{x}_j))}{\sum_{k \in N(i)} \exp(\text{MLP}(\mathbf{x}_i, \mathbf{x}_k))} \mathbf{W} \mathbf{x}_j,$$

where $\mathbf{x}_i$ and $\mathbf{x}_j$ are node features for nodes $i$ and $j$, respectively, node $j$ is taken from the neighbor $N(i)$ of node $i$, and $\mathbf{W}$ is a learnable matrix for processing node features.

While the attention mechanism is similar to that of Transformers for sequences, the advantage of using GATs is that node adjacencies themselves serve as important indicators of node dependencies. For sequences, other than by using proximity, it may be difficult to create these connections.

### 2.3.5 GIN

While the previous GNN architectures can be applied to polymer graph property prediction, they are primarily utilized for node-level tasks, in which individual node properties are typically predicted. Additionally, a limitation of both GCNs and GraphSAGE is that of injectivity. Injectivity is a property of functions that asserts that different inputs should be mapped to different outputs. GCN and GraphSAGE cannot act as injective functions because their aggregation functions are unable to distinguish neighborhoods with the same proportion of elements. For example, say that an atom within a polymer has two carbon neighbors and another atom has four carbon neighbors. Let us assume that each of these carbon neighbors only has one other neighbor. Using a GCN convolution, the atom with two carbon neighbors and the atom with four carbon neighbors are indistinguishable from each other, as the aggregated values of the two carbon neighbors are the same as the aggregated values of the four carbon neighbors. Similarly using GraphSAGE's sampling method, two

## 2.3 Graph-Based Prediction

carbon atom neighbors sampled from the second atom will cause the representation to be the exact same as the first atom with only two carbon neighbors. The maximum aggregation function for GraphSAGE is also inherently not injective. A new type of GNN, the Graph Isomorphism Networks (GINs) (Xu et al. 2019), addresses the injectivity issue of GCNs and GraphSAGE and are designed to solve graph-level tasks more specifically.

GINs are able to be injective functions because of the use of multi-layer perceptrons (MLPs) as aggregating functions. GINs also add an additional learnable parameter to scale a node's embedding during the combination step after aggregation. We can describe the aggregation and update function of GINs using the following equation:

$$\mathbf{h}_i = \text{MLP}((1 + \epsilon) \cdot \mathbf{x}_i + \sum_{j \in N(i)} \mathbf{x}_j),$$

where node features $\mathbf{x}_i$ for node $i$ and aggregated node features $\mathbf{x}_j$ for all nodes $j$ in neighborhood $N(i)$ are summed and passed through an MLP to generate the new node embedding $\mathbf{h}_i$.

The "isomorphism" in GIN refers to a mathematical property in which there exists a structure-preserving mapping between two structures. For graph isomorphism, this implies that for two graphs $G$ and $H$, there is a mapping $f$ between the graphs' associated sets of nodes such that two nodes $u$ and $v$ in $G$ are only adjacent if $f(u)$ and $f(v)$ are adjacent in $H$. Graph isomorphism is a difficult problem to be solved algorithmically, and it is debated whether this problem belongs in the set of problems known as NP-complete, difficult decision problems whose only solution may be to use brute force.

GINs, while not having the discriminative power to determine graph isomorphism exactly, are as powerful as a graph isomorphism heuristic called the Weisfeiler-Lehman (WL) graph isomorphism test (Weisfeiler and Leman 1968). If two graphs cannot pass the WL-test, then we can say for certain that they are not isomorphic. This does not imply that all pairs of graphs that pass the test are indeed isomorphic, but the heuristic is still useful to eliminate certain candidates. GINs are able to approximate the WL-test, meaning that if two graphs are given different embeddings using a GIN, then the WL-test will also determine that these graphs are not isomorphic. This discriminative power of GINs has made it very useful for graph prediction tests, including polymer prediction.

### 2.3.6 Graph Transformers

While GNNs have shown tremendous ability to represent graph data, they are inherently limited by what is known as graph over-squashing. During message passing, information from all of a node's neighbors are aggregated together to create a new node representation. However, over several layers of the GNN, information from previous layers may be lost. This issue is reminiscent of a similar problem in RNNs where information found early in a sequence may be lost over time. So, inspired by transformers used for text and other sequential data, we have Graph Transformer Networks, which similarly address information loss. The key to adapting transformers to graphs is the ability to leverage the attention mechanism to attend to graph structured data as opposed to sequential data. Just as with GATs, Graph Transformers leverage attention in the graph data, but complement full attention with local attention, rather than solely focusing on local neighborhoods. This means that long-distance dependencies of the graph can be captured directly rather than having to leverage multiple layers of attention. However, GAT's benefit over Graph Transformers when it comes to scalability. For large graphs, such as networks, calculating attention on a graph with thousands of nodes would be feasibly impossible, whereas attention on only local neighbors would relieve the heavy calculation load.

Transformers for sequences add positional encodings to each of its inputs to provide the attention mechanism with knowledge of the sequential relationships within the data. Graphs, however, do not have a natural and regular ordering of nodes as do items in a sequence. To provide positional knowledge of the nodes in the graph, we use Laplacian Positional Encodings (Belkin and Niyogi 2003). These PEs are created using the eigenvectors of a Laplacian matrix of a graph. The Laplacian PEs can be seen as an extension of the sinusoidal PEs used in the standard transformer as the Laplacian PEs of a $1 - D$ line graph resemble sinusoidal PEs.

The architecture of Graph Transformer layers resembles that of standard Transformers. The inputs to the Graph Transformer are first processed to generate initial node embeddings, which are concatenated with Laplacian PEs. The input embeddings are then passed through a series of encoders, composed of a multi-head self-attention layer and a feed-forward network. Each of these layers also adds residuals and normalizes the sum, similar to the standard Transformer.

## 2.4 Specific Techniques

In the previous sections, we have discussed important neural network architectures that can be used for polymer property prediction and their strengths. However, creating more powerful neural networks is not the only way to improve predictive performance. In this section, we will discuss these specific techniques, including optimizing model training by using hyperparameter tuning, solving the small data problem by using data augmentation, and a method to utilize both previous model training and new data by leveraging new learning paradigms.

### 2.4.1 Hyperparameter Tuning

One way to improve the performance of our models is by investigating the model's hyperparameters, which are variables of a neural network that are set prior to training a model by

## 2.4 Specific Techniques

its user. We can tune the values of these hyperparameters to those that allow the model to achieve better performance. Hyperparameter tuning does not affect the architecture of the model in terms of data flow, but it may impact the size and number of layers. This type of hyperparameter is known as the model hyperparameter. A second type of hyperparameter is used to affect the training of the neural network, known as algorithm hyperparameters. These hyperparameters will determine the behavior of the learning algorithm.

One type of model hyperparameter is the number of layers in an RNN, GNN, or (Graph) Transformer. Generally, too few layers prevents the model from learning more complex patterns in the data. However, too many layers may cause the model to overfit on the training data and negatively affect the performance on testing data. Additionally, too many layers in message-passing graph neural networks may cause what is known as over-smoothing. Signals from nodes in a graph can over time converge to the same value, causing this over-smoothing effect. Hence, balancing the number of layers is important to model performance. Another model hyperparameter is the latent dimension of the model. When we pass features into the first layer of the neural network, we are given an output vector of values, known as the latent representation. We have control over the size of the outputting vector, which is the latent dimension. Smaller latent dimensionality is quicker to process but may cause the representation to lose feature knowledge. Larger dimensionality will prevent this knowledge loss but is much slower to process. Too large latent size can also mean that some dimensions do not contain any useful knowledge at all. On a related note, in any feedforward network or MLP used by a neural network, we can also decide the dimensionality of its hidden layers. One last hyperparameter is that of batch size. In practice, is it infeasible to process all the data at the same time, as this would include thousands/millions of examples. We can split the data in smaller groupings called batches, and these batches are processed together.

Algorithm hyperparameters are those that affect one of the key components of polymer property prediction, that being model optimization. A standard optimization algorithm is gradient descent, which will try to minimize the error of the model during training in order to create the best predictions during testing. The error to minimize is calculated using a loss function or objective function during several training repetitions, and model weights can then be updated with this error, or loss, in mind. We mention gradient descent, which calculates the first derivative or gradient of the loss function and "descends" to a local minimum of the function, but there are actually several different optimization algorithms we can choose, such as Adagrad, RMSProp, or Adam. Furthermore, with the direction of the optimization determined by the type of optimization algorithm, we can also adjust the speed of optimization, using a term called a learning rate. A large learning rate will make large jumps and speed up optimization. However, it may take longer to converge to the local minimum with these larger adjustments. A smaller learning rate may take many more training iterations but should converge more stably. The learning rate may also inform the number of training iterations or epochs we will train the model. Fewer epochs may not allow the optimization to converge, while greater epochs may cause the model to overfit on the training data. We can also choose the loss function for the learning. We discussed in a previous section the different metrics for classification and regression tasks, but in theory we could train with any objective in mind. One last hyperparameter to consider is the activation function in the neural network layer. We mentioned ReLU in a previous section, but other activations include Sigmoid, Tanh, and Leaky ReLU.

### 2.4.2 Data Augmentation

Data augmentation is a strategy for increasing the number of training examples without needing to gather and label more data, a process which can be time-consuming and expensive. Data augmentation will take a sample-label pair and modify the sample slightly. The new modified sample will resemble the original, and we will use the same label for this new sample. For a simple example, we may look at different ways to perform data augmentation on images. If we wanted more pictures of dogs to train an animal classifier, we could flip, rotate, crop, stretch, warp, or recolor an image of a dog. These will all still be images of a dog, but the model now has many more images to use in training. We can similarly augment polymer data to increase the number of samples we have for training. If hyperparameter tuning is a way to improve our model by refining the learning process, then data augmentation is a way to improve our model by extracting more knowledge out of our available training data. In the next sections, we will discuss both SMILES-based and graph-based augmentation strategies.

#### 2.4.2.1 SMILES-Based

SMILES-based augmentations can occur in one of two ways, by modifying the SMILES itself or the associated label value. There are several ways to perform SMILES augmentation. One way is to use non-canonical forms of a molecule. We can essentially reorder the sequence of the SMILES so that even though the string represents the same polymer, the exact sequence is different. This method is useful as the chemical validity and label integrity are preserved. We may also use a generative model to create a SMILES string given a specific label value. Generative models will be discussed in a later section. The second type of augmentation is based on the SMILES label value. If the data we have is missing label values, one strategy is to first train a model to create label prediction or pseudo-labels for the data. These new pseudo-labeled samples can then be used to improve the training of the prediction model. Note that the pseudo-labels can be progressively updated as the model performance improves.

#### 2.4.2.2 Graph-Based

While SMILES-based data augmentation is a valid way to increase the number of training samples, it is less commonly used as compared to graph-based augmentation strategies due to the difficulty in creating valid new SMILES, both in terms of syntax and chemical validity. The existing strategies for graph-based data augmentation can be broadly separated

## 2.4 Specific Techniques

into three types, whereby a graph is augmented by changing either features, structure, or labels. By perturbing any one of these aspects of graph data, we will be able to generate new labeled graphs to use in training. Starting with feature augmentation, we may change the node features of the graph. We can add noise to the original node features in a process called feature corruption. We could additionally mask specific features or add new ones. By modifying the feature values of the graph nodes, we can create new graphs with the same structure and labels. If we choose to augment structure, instead of modifying nodes, we modify the set of edges in the graph. Using edge perturbation, we can randomly add or remove edges from the graph structure to create new graphs. Graph rewiring is similar to edge perturbation, but the edge addition or removal is guided by the downstream predictive task with the goal of making the graph data more useful. We may also choose to add or remove nodes from the graph to create augmented examples. Another way is to perform (sub)graph sampling, in which augmented samples are taken from a subset of nodes and edges from the original graph. A final structure-oriented augmentation strategy is graph generation which will be discussed in later sections. Put briefly, a graph generation model can be trained to create samples that closely align with the structures of real data. Finally, we have different label-oriented augmentation strategies. As explained in the previous section, we can generate pseudo-labels for unlabeled graph samples. The second strategy is to use label mixup, which is a hybrid between feature augmentation and label augmentation. To generate a new sample, we take a weighted average of the features and label values between two samples. As the new sample resembles both of the original samples, the new label value will most likely be a value in between the labels of the originals as well.

### 2.4.3 Other Learning Paradigms

The previous strategies discussed looked at both the data and the model training as avenues for improving model performance. This next method is, in a way, a combination of both, where we can utilize more data and further finetune model weights during training. The type of model training we have discussed previously involves using labeled training data to predict the values of test data in which the training data and test data come from the same larger dataset. This is known as supervised learning, and it is one of the most common learning strategies. However, there are other learning paradigms that train using data differently. Transfer learning is another training paradigm where knowledge from one training task is reused for another. If a model is trained on more than one dataset and the model weights are shared for several supervised tasks, this is known as multi-task learning. If the model is trained without using the training labels, this is known as self-supervised learning. These three training paradigms, which we visualize in Fig. 2.4, allow for models to exploit knowledge that may not be considered in a supervised framework.

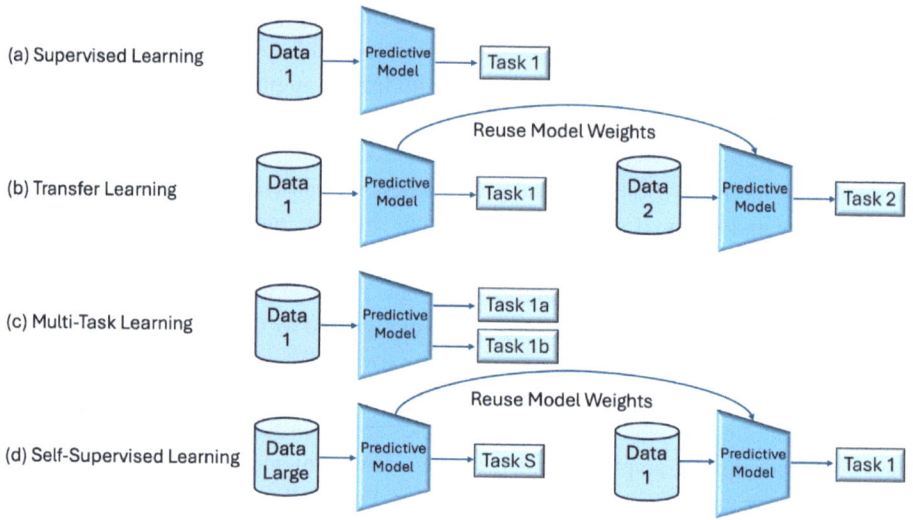

**Fig. 2.4** Visualizations of different learning paradigms, supervised (**a**), transfer (**b**), multi-task (**c**), and self-supervised learning (**d**). Transfer learning reuses model weights for a second task. Multi-task learning trains a model for two tasks on the same dataset. Self-supervised learning initially trains a model on a self-supervised task S on a large unlabeled dataset before training on the test task

### 2.4.3.1 Transfer Learning

The first alternative learning strategy we will discuss is transfer learning, or domain adaptation. It is motivated by the idea that different tasks in the same domain may be similar and the knowledge learned from the model on one task will benefit a model on another task. This paradigm is meant to resemble human learning, where our existing knowledge base can be used for new tasks. Given two tasks where the input data is from the same domain, then we can train a supervised model on the first task and later retrain the model on the second task. The model weights learned from the first task transfer over relevant domain knowledge that would be absent using a random weight initialization.

### 2.4.3.2 Multi-task Learning

Transfer learning trains several tasks in sequence. However, we may consider training several tasks in parallel, which is called multi-task learning. This paradigm aims to help models generalize their learning and allow knowledge from different tasks to be accumulated and positively influence the model. Multi-task learning resembles human learning as training several skills together can mutually benefit performance. Polymers may have two or more properties, and we can train one model to predict all of these properties together rather than train many models. There are two different kinds of multi-task learning, depending on the way model weights are shared between the tasks. Hard parameter sharing will use shared neural network layers and add a single final task-specific layer at the end of the pipeline for each task. Soft parameter sharing separates each task into its own network and pipeline, but the weights are regularized across layers so that the models' weights are close in distance.

### 2.4.3.3 Self-supervised Learning

The last learning paradigm we will discuss is self-supervised learning, which assumes that relevant knowledge is contained within the data itself without need for explicit training labels. This paradigm is often associated with model pre-training, where a model's weights are "warmed-up" on a self-supervised task first, before using the model for a supervised learning task. Often, the dataset used for the self-supervised task is much larger than the task-specific dataset as the non-necessity of labels allows for the collection of a greater number of samples. In a way, self-supervised learning can be seen as a type of transfer learning, where the objective of the initial training task is determined by implicit labels of the data. Model pre-training for polymer prediction generally falls into two categories: predictive and contrastive. Predictive tasks involve predicting or reconstructing some feature of the data. For polymer property prediction models, masked language/attribute modeling is often used to pre-train models. The goal of masked modeling is to predict the value of a language token or graph node that has been removed from the data. For example, in a SMILES string or polymer graph, we can mask the atom type of a carbon atom and let the model try to predict what this element should be based on the rest of the data. Contrastive model pre-training will compare the similarity of different data samples and try to preserve this similarity within the latent representation space. Polyethylene and polypropylene, for example, are structurally very similar, with a single hydrogen being replaced with a methyl group, so these two polymers will be mapped to similar latent space representations. In contrast, polyamide will be mapped to a very different latent space representation as it is very dissimilar from polyethylene, containing amide groups.

## 2.5 Summary

Polymer property prediction is an important task in polymer informatics with the goal of inputting polymer data into a predictive neural network and generating label values for that data sample. The formal definition of the prediction task is guided by the type of labels we aim to predict. These labels can be categorical, leading to a classification task, or they could be continuous, which turns the task into a regression. Whether the task is the classification of regression, the way to perform property prediction is by utilizing neural networks, which can be designed around taking either sequential SMILES data or graph data as input. Using a sequential representation of polymers, we can utilize neural networks called RNNs which are specifically built to understand the positional relationship between elements of the input. Modifications such as LSTMs and GRUs aim to bolster the capabilities of RNNs by implementing a type of "memory" mechanism. Given some of the inefficiencies associated with RNNS, we may instead choose to use Transformers to process our sequential data, which use an attention mechanism to learn dependencies within the data. Using the more popular graph representation for polymers, we can utilize GNNs to process polymer data. These GNNs, such as GCNs, GraphSAGE, GATs, and GINs, all follow a local neighborhood feature aggregation and combination strategy. To address limitations of GNNs such as over-squashing, Graph Transformers, inspired by classical sequence Transformers, were proposed to bring a global attention mechanism to graph data. Given all of these different neural network architectures for processing polymer data, we can use a few additional techniques to further improve predictive performance and determine which model is most useful for certain applications. We can optimize the neural network itself by tuning hyperparameters that guide learning. We can also create extra data via data augmentation to increase the number of training samples our model can learn using. A final method, which involves training models using different learning paradigms, is able to both finetune the model weights during training and increase the number of training samples. With these methods and the discussed neural network architectures, it is possible to create powerful prediction models for polymer informatics.

## References

J. Achiam, S. Adler, S. Agarwal, L. Ahmad, I. Akkaya, F. L. Aleman, D. Almeida, J. Altenschmidt, S. Altman, S. Anadkat, et al. Gpt-4 technical report. *arXiv preprint* arXiv:2303.08774 2023.

M. Belkin and P. Niyogi. Laplacian eigenmaps for dimensionality reduction and data representation. *Neural computation*, 15(6):1373–1396, 2003.

A. Chandrasekaran, C. Kim, S. Venkatram, and R. Ramprasad. A deep learning solvent-selection paradigm powered by a massive solvent/nonsolvent database for polymers. *Macromolecules*, 53(12):4764–4769, 2020.

J. Devlin, M.-W. Chang, K. Lee, and K. Toutanova. Bert: Pre-training of deep bidirectional transformers for language understanding. *arXiv preprint* arXiv:1810.04805 2018.

W. L. Hamilton, R. Ying, and J. Leskovec. Inductive representation learning on large graphs. In *Proceedings of the 31st International Conference on Neural Information Processing Systems*, pages 1025–1035, 2017.

T. N. Kipf and M. Welling. Semi-supervised classification with graph convolutional networks. In *International Conference on Learning Representations*, 2017.

C. Kuenneth and R. Ramprasad. polybert: a chemical language model to enable fully machine-driven ultrafast polymer informatics. *Nature Communications*, 14(1):4099, 2023.

R. Kumar, N. Le, Z. Tan, M. E. Brown, S. Jiang, and T. M. Reineke. Efficient polymer-mediated delivery of gene-editing ribonucleoprotein payloads through combinatorial design, parallelized experimentation, and machine learning. *ACS nano*, 14(12):17626–17639, 2020.

S. Otsuka, I. Kuwajima, J. Hosoya, Y. Xu, and M. Yamazaki. Polyinfo: Polymer database for polymeric materials design. In *2011 International Conference on Emerging Intelligent Data and Web Technologies*, pages 22–29. IEEE, 2011.

M. G. Project. Polymer property predictor and database. URL https://pppdb.uchicago.edu/.

D. Rouvray. The pioneering contributions of cayley and sylvester to the mathematical description of chemical structure. *Journal of Molecular Structure: THEOCHEM*, 185:1–14, 1989.

# References

A. Thornton, L. Robeson, B. Freeman, and D. Uhlmann. Polymer gas separation membrane database, 2012. URL https://research.csiro.au/virtualscreening/membrane-database-polymer-gas-separation-membranes/.

A. Vaswani, N. Shazeer, N. Parmar, J. Uszkoreit, L. Jones, A. N. Gomez, Ł. Kaiser, and I. Polosukhin. Attention is all you need. *Advances in neural information processing systems*, 30, 2017.

S. Wang, Y. Guo, Y. Wang, H. Sun, and J. Huang. Smiles-bert: large scale unsupervised pre-training for molecular property prediction. In *Proceedings of the 10th ACM international conference on bioinformatics, computational biology and health informatics*, pages 429–436, 2019.

B. Weisfeiler and A. Leman. The reduction of a graph to canonical form and the algebra which appears therein. *nti, Series*, 2(9):12–16, 1968.

K. Xu, W. Hu, J. Leskovec, and S. Jegelka. How powerful are graph neural networks? In *International Conference on Learning Representations*, 2019. URL https://openreview.net/forum?id=ryGs6iA5Km.

# Deep Learning for Inverse Polymer Design

## 3.1 Problem Definition and Datasets

Neural networks can be utilized to generate polymers either randomly or according to specific human-defined requirements. These requirements can be represented as categorical data, numerical values, or text. Depending on the neural network architecture, polymer structures can be generated sequentially (using Recurrent Neural Networks, RNNs) or in a one-shot manner (using Graph Neural Networks, GNNs). In this chapter, we will first introduce the problem of polymer inverse design with a specific example related to gas separation membrane designs. Following this, we will explore popular neural network architectures for polymer generation, addressing two common generation tasks: unconstrained and constrained generation.

### 3.1.1 Unconditional Generation Without Constraints

Unconditional generation refers to the process of creating new data, such as polymer structures, without any specific requirements or constraints. In this process, the model inputs could be random noise from the normal distribution. The model generates outputs based solely on the patterns it has learned from the training polymers, without any additional input or guidance. This type of generation allows for the exploration of a wide variety of possible outputs, which can be useful for discovering new and innovative structures.

Suppose we aim to design polymeric membranes with high oxygen permeability for new face masks. We can start by collecting a batch of polymers known for their high oxygen permeability. These polymers are used to train a generative model. The model maps these polymers into a latent space, a lower-dimensional representation that captures the essential features of the polymers. By exploring this latent space and randomly sampling points within it, the model can generate new polymers, which can then be examined for novel applications.

### 3.1.2 Conditional Generation With Constraints

Unconditional generation is impractical when designing polymers with specific properties. In this case, the generation model takes the specific property label or constraint $y$ as input. The model generates samples that satisfy the given condition, ensuring that the new data point x aligns with the associated label $y$.

Practical polymers must meet various requirements, such as strength, flexibility, conductivity, and thermal stability. For gas separation applications, polymers need specific permeabilities to different kinds of gas. For example, to reduce carbon emissions, a polymer might require high oxygen permeability but low carbon dioxide permeability. This enables polymeric membranes to separate these gasses, capturing carbon before it is emitted.

To meet multiple property requirements, constraints can be represented as a vector, with each element corresponding to a different property. This vector guides the polymer generation process by serving as a set of guidance signals for the neural networks, enabling them to produce polymers with the desired characteristics. The specific usage of these guidance signals depends on the type of neural network architecture used.

Generally, we can encode these multi-property requirements into the latent space of the model. By interacting with these latent representations, we can update the latent vectors of the polymers. Interaction methods include simple concatenation of the vectors, adding them together, or using more sophisticated approaches like the cross-attention mechanism. These techniques allow the model to integrate multiple property constraints effectively, ensuring the generated polymers meet the specified requirements.

### 3.1.3 Datasets, Task Formulation and Evaluation

The datasets used for polymer inverse design are often the same as those used for property predictions, consisting of numerous pairs of polymer structures and their corresponding property labels. Unlike property prediction models, polymer inverse design uses property labels as inputs to generate polymer structures that match these properties. Additionally, it is desirable for the generated structures to resemble the target structure while maintaining a certain level of diversity.

Polymers with annotated properties are challenging to collect due to the lengthy and expensive nature of chemical experiments. However, online resources like PolyInfo (Otsuka et al. 2011) and MSA (Thornton et al. 2012) provide relatively large-scale datasets of labeled polymers. Polymers can be represented using the p-SMILES format. For example, poly(ethylene terephthalate), commonly used in fibers for clothing and containers for liquids and foods, is represented as "*OCCOC(=O)c1ccc(C(=O)O*)cc1," where the SMILES string denotes the monomer structure, and asterisks indicate polymerization points.

Denoting the polymer structure as $x$ and its properties (e.g., permeabilities for various gases like $H_2$, $O_2$, $CO_2$, $CH_4$) (Barnett et al. 2020) as a vector $y$, the inverse design task aims to design a neural network-based generation model $g(y)$ to construct the polymer $x'$. In some cases, $y$ may not be provided, resulting in an unconstrained model. The evaluation objective is to minimize $\text{sim}(x', x)$, a similarity evaluation metric for polymer structures, and maximize $O(x', y)$, an oracle evaluation metric for properties, when $y$ is available.

Various metrics can be used for evaluating $\text{sim}(x', x)$. Point-wise metrics measure the similarity between individual pairs of $(x', x)$, such as fingerprint cosine similarities. Distribution-based metrics compare the overall similarity between the set of generated polymers and a reference set. For instance, BRICS-based fragments (Degen et al. 2008) and scaffolds (Bemis and Murcko 1996) can identify substructures in both the reference and generated sets to compute their similarity.

It is crucial to consider diversity metrics for the group of generated polymers to ensure the model does not simply repeat the same results. Two key metrics are Novelty and Uniqueness. Novelty measures the ratio of generated polymers not present in the training set, while Uniqueness measures the ratio of non-repeating generated polymers. However, these instance-level metrics do not fully capture the diversity of the generated set. Interval diversity Chen et al. (2018), which measures the fingerprint similarities between pairs in the generated set, provides a better reflection of how diverse the generated polymers are.

Accurate property evaluation through molecular dynamics simulations and chemical experiments can take weeks or months, making them impractical for evaluations of polymer generation models. Typically, we could use well-trained prediction models as proxies for the oracle function in property evaluation (Gao et al. 2022). We can train a neural network model as the property evaluator. Any neural network-based evaluation functions, as discussed in Chap. 2, can be used. Since predictor models may not provide 100% accurate evaluations, using multiple prediction evaluators and considering the variance across different evaluations can improve reliability.

In the following sections, we will introduce several popular neural network models for $g(y)$ and discuss their specific applications in both unconstrained and constrained polymer generation.

## 3.2 Generative Neural Network Architectures

In this section, we introduce three popular neural network architectures for polymer generation: Generative Adversarial Networks (GANs), Variational Autoencoders (VAEs), and Diffusion Models (DMs). Their comparisons are shown in Fig. 3.1.

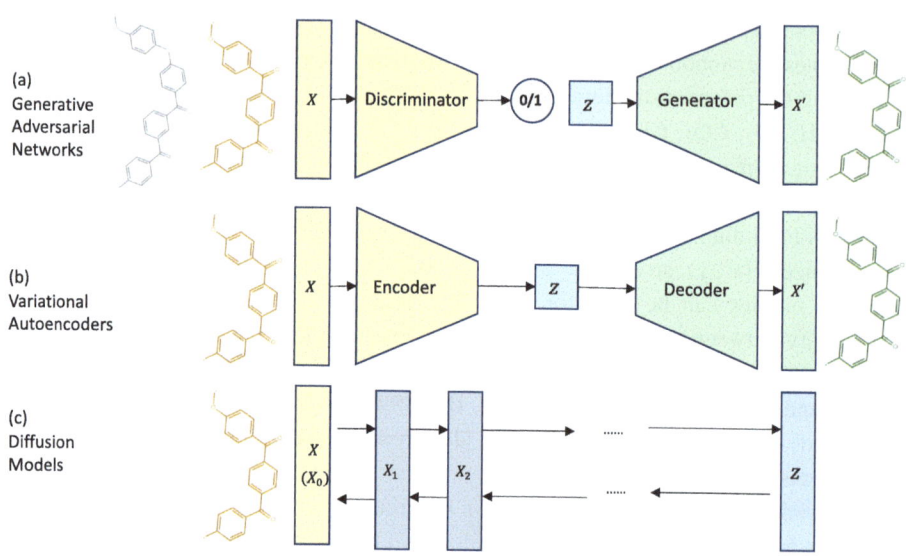

**Fig. 3.1** Visualization of different generative neural network architectures: **a** generative adversarial networks; **b** variational autoencoder; **c** diffusion model

## 3.2.1 Generative Adversarial Networks

Generative Adversarial Networks (GANs) are a type of generative model that implicitly estimates probability density or mass function of a distribution (Goodfellow et al. 2020). They consist of two neural network submodules: the generator and the discriminator. The generator takes random noise and additional property vectors as input and proposes polymer structures. The discriminator then tries to distinguish these generated structures from real polymers. The generator's goal is to create polymers that are indistinguishable from real ones, while the discriminator aims to accurately identify the generated polymers. These two networks are jointly optimized, and the process reaches equilibrium when the generator successfully fools the discriminator, making it unable to distinguish between real and generated polymers. Eventually, the generator can propose new polymers similar to the training data. It is termed 'adversarial' because the two sub-networks engage in a min-max game: the generator maximizes the likelihood of the generated polymers resembling real ones, while the discriminator minimizes this likelihood.

## 3.2.2 Variational Autoencoders

GANs implicitly model the probability density of distributions and are known to be difficult to train (Salimans et al. 2016). Is there a way to directly model the data distribution, such as

## 3.2 Generative Neural Network Architectures

in a latent space? The answer is Autoencoders (AEs). It is another basic neural architectures for generating data. They compress input polymers into a low-dimensional vector space and then recover the input features from this latent space. AEs consist of two parts: an encoder network, which maps the input to the latent space, and a decoder network, which maps the latent space back to the reconstructed inputs. Both the encoder and decoder are jointly trained. By randomly sampling data points in the latent space, the decoder can generate new polymers. However, random generation is not sufficient for designing new polymers with specific properties. While AEs minimize the reconstruction loss of polymer structures, they do not ensure that the latent space is well-organized or regular. This means that although AEs are good at compressing polymer structures, the lack of regularity in the latent space makes controlled generation challenging.

Variational Autoencoders (VAEs) address this issue by regularizing the latent space in addition to minimizing reconstruction losses (Kingma 2013). VAEs can be thought of as regularized AEs that better organize the latent space for polymer generation. Unlike AEs, VAEs encode input polymers into a latent distribution rather than a single point. This allows sampling from the distribution to control the generation process. The training process for VAEs involves three steps: encoding the polymer into a distribution, sampling a data point from the encoded distribution, and decoding the sample to reconstruct the polymer. The latent distribution is usually chosen to be a normal distribution. During model training, the input polymers are encoded into a mean and covariance matrix that describe this distribution, facilitating sampling. To construct such a distribution, the model minimizes the distance between the learned distribution and the prior distribution, often using the Kullback-Leibler (KL) divergence as a measure of this distance. This regularization ensures that the latent space is well-organized, enabling the generation of polymers that meet specific requirements.

### 3.2.3 Diffusion Models

While VAEs model data distributions in latent space, diffusion models offer an alternative approach by gradually transforming simple distributions into complex data through iterative processes. Diffusion models, also known as diffusion probabilistic models or score-based generative models, are inspired by non-equilibrium thermodynamics. They define two processes: forward and reverse diffusion. In the forward diffusion process, each step adds small noise, typically sampled from a Gaussian distribution, to the data. The reverse diffusion process, also known as the data generation process, starts from randomly sampled noise and uses a neural network to gradually remove the noise, ultimately generating new polymers.

Both forward and reverse diffusion are defined as Markov processes, meaning each step depends only on the state from the previous step. Suppose there are $T$ diffusion steps, typically set to 500 or 1000. The forward diffusion process $p(X_{1:T}|X_0)$ is usually predefined, where $X_0$ is the input data point such as molecules, polymers, or images. $X_1, X_2, \ldots, X_T$ denote the noisy states of the input data points at diffusion steps $1, 2, \ldots,$ and $T$.

Initially, diffusion models such as DALL-E (Ramesh et al. 2021) focused on images, adding Gaussian noise while maintaining the same dimensions as the input images. To enhance the effectiveness and efficiency of diffusion models, various approaches explore diffusion processes in discrete structure spaces (Vignac et al. 2022) or latent spaces (Rombach et al. 2022). Noise scheduling is another important technique in diffusion models, controlling the amount of noise added between the current step $t$ and the next step $t+1$. Too much noise makes the neural network's estimation difficult, while too little noise causes inefficiency.

The training goal of the diffusion model is to develop a denoising model or denoiser that accurately estimates the noise in the reverse diffusion process $p(X_{t-1}|X_t)$. This enables the removal of noise at the current diffusion step, gradually progressing toward the real data distribution. Noise estimation can be categorized into three types, each resulting in different training losses:

1. Estimating Noise Between Adjacent Steps (Ho et al. 2020): This involves estimating the noise $e$ between two adjacent steps and inferring the state at step $t-1$, i.e., $p(X_{t-1}|e)p(e|X_t)$. The noise in this category is often continuous, such as noise from a normal distribution. The mean squared error (MSE) is commonly used as the loss function.
2. Estimating the Denoised Data Point Directly: This method involves estimating the denoised data point directly, $p(X_{t-1}|X_t)$. The training loss depends on the type of data. For images, the MSE loss function is used. For polymers and small molecules, the model predicts the types of atoms and bonds, and the cross-entropy (CE) loss is used for training.
3. Estimating the State at $X_0$ (Austin et al. 2021): This method estimates the state at $X_0$ and then uses the forward diffusion to infer $X_{t-1}$, i.e., $p(X_{t-1}|X_0)p(X_0|X_t)$. Similar to the second category, for polymers and small molecules, the model predicts the types of atoms and bonds, and the CE loss is used.

## 3.3 Unconstrained Polymer Generation

In this chapter, we explore the use of generative models in unconstrained polymer generation, where polymers are randomly generated within the training data distribution. Polymers can be represented as sequences (e.g., by SMILES strings) or as graphs. An overview of the generation procedures for sequences and graphs is shown in Fig. 3.2. Depending on the chosen representation, we can select appropriate neural network architectures such as Long Short-Term Memory (LSTM) and Graph Convolutional Network (GCN), as discussed in Chap. 2, for these generative models.

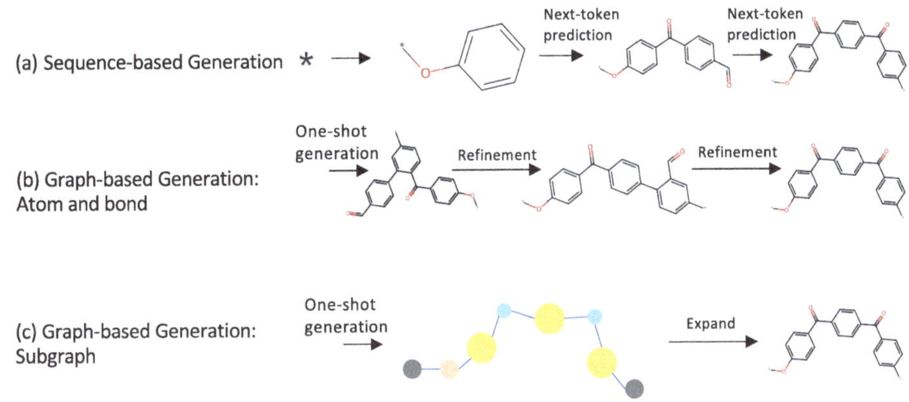

## 3.3 Unconstrained Polymer Generation

**Fig. 3.2** Comparison of sequence and graph-based generation procedures: **a** sequence-based next token prediction; **b** graph-based atom and bond generation and refinement; **c** graph-based subgraph generation

### 3.3.1 Sequence-Based Generation

Sequence-based generation creates the basic units of sequential polymer SMILES strings, known as tokens, step-by-step. This process first requires tokenization, which builds a dictionary mapping unit symbols to vectors. After converting the input into vectors, the sequential model uses neural network architectures to learn which token to decode at each step. After training, we can sample new polymers by generating tokens sequentially until a special stop token is triggered.

#### 3.3.1.1 Tokenization

This step splits the polymer sequence into smaller units, creating a dictionary with token symbols as keys and embedding vectors as values. Each token's embedding can be retrieved from this dictionary, and the similarity between the output embedding and the dictionary embeddings can be calculated to decode the token symbols. Tokenization does not necessarily ensure that polymer units are atom by atom.

In natural language processing, common tokenization techniques include Byte-Pair Encoding (BPE). When applied to polymer SMILES strings, BPE starts with the most basic elements in the string, such as atom types (e.g., 'C', 'N', 'O', [B], [Si]), bond types (e.g., single bond '=' and triple bond '#'), digits indicating ring formation, chirality, etc. These basic units initialize the token dictionary.

BPE then iteratively updates the dictionary with frequent unit pairs. It identifies the most frequent pairs among the units (according to a corpus, such as a large collection of polymer SMILES) and adds the most frequent one as a new token to the dictionary. This process repeats until a specified number of iterations is reached or the minimum frequency threshold is met.

Tokenization defines a dictionary used for polymer encoding and decoding, making it impossible to generate polymers with symbols or atoms not included in this dictionary. Typically, we create the dictionary by applying tokenization methods to a pre-collected polymer SMILES corpus. Consequently, atoms that are rare in common polymer corpus, if not included in the dictionary, cannot be designed. To facilitate generation, for example, we can insert special tokens into the dictionary to indicate when to stop the polymer generation.

### 3.3.1.2 Choice of Architecture

There are many neural network models and generation architectures available for polymer generation, including LSTM and Transformer models introduced in Chap. 2, as well as specialized architectures like GAN and VAE. To generate polymer SMILES, these models must be adapted in an autoregressive manner. In each step of autoregressive generation, the models take the current state as input, summarize information from previous states, and predict the next token. We refer to these models in their autoregressive form as autoregressive models.

Autoregressive models need to properly integrate previous and current states to predict the next token. For example, LSTMs dynamically update hidden states to summarize previous states, while Transformers use causal attention to calculate dot product similarities and summarize previous states into the current one. By stacking multiple layers of neural networks, autoregressive models output a vector with the same dimensions as the token vectors in the dictionary. The cosine similarity between the output vector and all dictionary vectors is computed, and the most similar token is selected for the next step. To generate a polymer, the process starts with a special token indicating the starting point and continues until another special token indicates stopping.

To enable long-term generation, LSTM and Transformer models are trained using cross-entropy loss for prediction. Their generation ability can be further improved by integrating these models into GAN and VAE architectures. In GANs, incorporating LSTM or Transformer models into the generator allows leveraging the discriminator to evaluate the authenticity of generated polymers by distinguishing them from real ones. Through adversarial training, the generator learns to produce increasingly realistic polymer sequences, enhancing the quality and diversity of the generated polymers. In VAEs, incorporating LSTM or Transformer models enhances the encoding and decoding processes for sequential data. The encoder maps current and previous states into a latent space, while the decoder predicts the next token from this latent representation. This probabilistic framework allows for better modeling of the latent space for previous, current, and next states. Training these advanced models involves optimizing multiple objectives, such as reconstruction loss, adversarial loss, and KL divergence, depending on the architecture.

## 3.3.2 Graph-Based Generation

Graphs are more natural representations of polymers. While one polymer can correspond to multiple SMILES sequences depending on the initialization of the depth-first traversal in the SMILES algorithm, it has only one type of graph representation. Graph neural networks (GNNs) can be used as the basic encoding and decoding modules within architectures such as GANs, VAEs, and diffusion models to generate polymers. These models predict atom types, bond types, and bond connections, ultimately producing polymer graphs. Besides, generation models could be defined on subgraphs such as functional groups and motifs (frequent subgraph patterns) which enable more efficient and controllable polymer generation.

### 3.3.2.1 Generation Based on Atoms and Bonds

Polymers can be generated by predicting atoms (nodes) and bonds (edges) using two matrices: the node feature matrix and the adjacency matrix. In a polymer graph with $N$ nodes, the node feature matrix $\mathbf{X} \in \mathbb{R}^{N \times F_X}$ represents each atom type as a one-hot encoding, with each row corresponding to one of the $F_X$ atom types from the periodic table. The adjacency matrix $\mathbf{A} \in \mathbb{R}^{N \times N \times F_A}$ encodes bond types and their connections, with each row representing a bond type–single, double, triple, aromatic, or non-connection–so typically, $F_A = 5$.

To generate such matrices using a GAN, the generator takes a randomly sampled matrix as input, typically consisting of a few shared hidden layers and two linear transformation layers that map these hidden states into node feature and adjacency matrices. The GNN-based discriminator then predicts whether the polymers are generated or real. While more powerful GNNs can replace the MLP in the generator, caution is needed since message passing GNNs rely on local neighbors to update the central node representation. The generator works on messy adjacency matrices, which may negatively impact the message passing performance. To optimize the GAN model, we use min-max adversarial losses, where the generator is optimized to produce more realistic polymers and the discriminator is optimized to effectively distinguish between real and generated polymers.

In VAEs, GNNs can be used as encoders to model the latent space. This involves first modeling the joint distribution of $\mathbf{X}$ and $\mathbf{A}$ using shared GNN layers. From this, we construct two distributions by learning their mean and variance: one for node features and the other for the adjacency matrix. The VAE decoder module then takes randomly sampled node features and adjacency matrices from the latent distributions and uses another GNN model to reconstruct the input $\mathbf{X}$ and $\mathbf{A}$. The entire process is optimized using a reconstruction loss plus a KL divergence loss between the learned posterior distribution in the latent space and the prior normal distribution.

Diffusion models generate new polymers by gradually removing noise from randomly sampled $\mathbf{X}$ and $\mathbf{A}$. Here, $t$ denotes a diffusion step, and $T$ represents the total diffusion steps. The node feature and adjacency matrix at step $t$ are represented as $\mathbf{X}^t$ and $\mathbf{A}^t$, respectively. The noise can be defined in continuous or discrete spaces and must satisfy three criteria (Vignac et al. 2022): (1) The transition with noise must have a closed-form formulation, allowing us to directly obtain the noisy state of $\mathbf{X}$ or $\mathbf{A}$ at any step $t$, e.g., $p(\mathbf{X}^t|\mathbf{X})$ or $p(\mathbf{A}^t|\mathbf{A})$; (2) The posterior $p(\mathbf{X}^{t-1}|\mathbf{X}^t)$ or $p(\mathbf{A}^{t-1}|\mathbf{A}^t)$ should also have a closed-form expression to enable neural network denoising; (3) As diffusion steps approach infinity, the noisy states should converge to a prior distribution, allowing the use of random $\mathbf{X}$ and $\mathbf{A}$ from this prior distribution and gradual noise removal. Gaussian noise from a normal distribution typically satisfies these criteria for continuous noise (Jo et al. 2022). We add Gaussian noise separately to the node feature $\mathbf{X}$ and the adjacency matrix $\mathbf{A}$ and define a sufficiently large number of total diffusion steps $T$ to ensure $p(\mathbf{X}^T|\mathbf{X})$ and $p(\mathbf{A}^T|\mathbf{A})$ converge to the normal distribution. A neural network is trained to estimate the added noise from step $t-1$ to $t$, enabling the reverse diffusion process to remove noise from $t$ to $t-1$ until the original polymer is recovered.

Discrete noise can also be defined using a Markov transition matrix $\mathbf{Q} \in \mathbb{R}^{F \times F}$ ($\mathbf{Q}_X \in \mathbb{R}^{F_X \times F_X}$ for atoms and $\mathbf{Q}_A \in \mathbb{R}^{F_A \times F_A}$ for bonds) (Vignac et al. 2022). This method suits the discrete nature of polymer graphs, allowing precise definitions of transition probabilities between different atom and bond types. Atom and bond type frequencies are typically calculated from the training set and normalized to obtain probability vectors $\mathbf{m}_X$ and $\mathbf{m}_A$. These vectors are used to create the transition probability matrix $\mathbf{Q}$, which is often combined with an identity matrix and a weighting coefficient controlled by diffusion steps. As step $t$ decreases, the weight of the identity matrix increases, and vice versa. In diffusion models with discrete noise, we do not directly estimate matrix-based noise but predict atom and bond types directly. Using the closed-form forward diffusion process, we can infer state $t-1$ from $t$ through an intermediate approximation of $\mathbf{X}$ and $\mathbf{A}$. The denoising graph neural network is then optimized using the reconstruction loss from the noisy state $t$ to the original $\mathbf{X}$ and $\mathbf{A}$. In this case, the discrete diffusion model iteratively generates and refines the entire polymer structure in multiple steps.

### 3.3.2.2 Generation Based on Subgraphs

Graph structures offer a flexible approach to generating polymers subgraph-by-subgraph (Jin et al. 2018). Molecular substructures, such as scaffolds, functional groups, and motifs, are knowledge-intensive representations that can be intuitively assembled using graph structures. For example, we can extract a junction tree where each node represents a subset of atoms, turning the generation task into constructing trees. Within the VAE framework, we can design polymer encoders and decoders alongside a tree encoder-decoder. This approach results in two types of latent distributions for polymers and trees. During decoding, the generated tree can constrain the polymer structure, aiding in the design of polymers with desirable topologies.

Another advantage of subgraph-based generation is its ability to produce larger polymers. While we typically focus on single repeating units of polymers with polymerization points,

## 3.4 Constrained Polymer Generation

some polymers have larger single units with over a hundred atoms. Many atom- and bond-based generation models are developed and tested on smaller polymers (e.g., fewer than ten nodes) and become inefficient or run out of memory when scaling to larger ones (Jin et al. 2020). Subgraph-based generation treats molecular substructures as nodes, significantly reducing the size of the generated graphs and enabling the efficient creation of larger polymers.

## 3.4 Constrained Polymer Generation

In practical applications, researchers are often interested in a small group of polymers with specific desirable properties. These properties can be used as constraints to control polymer generation. As shown in Fig. 3.3, constraints can include categorical values, continuous values, polymer structures, and texts. This chapter discusses how to process these constraints and incorporate them into neural network frameworks, either as loss functions or input features, to guide polymer generation.

### 3.4.1 Constraint Types

#### 3.4.1.1 Multiple Properties

The most straightforward constraints come from property values. For example, in gas separation tasks, we might specify an oxygen permeability of 100 barrer and a $CO_2$ permeability

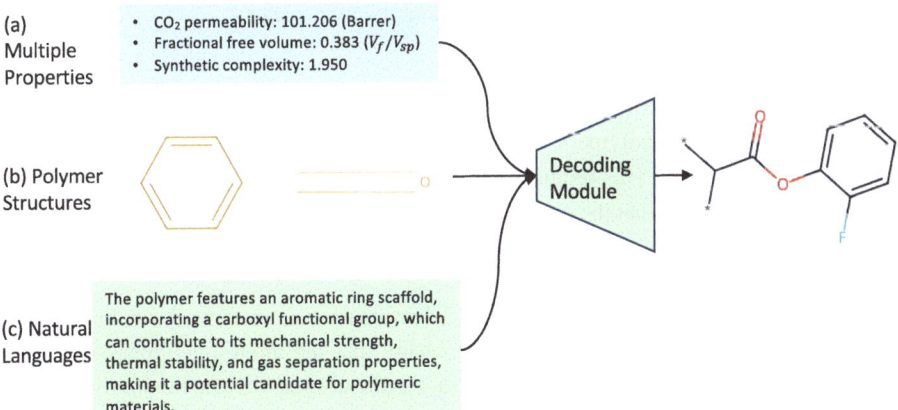

**Fig. 3.3** Different types of constraints for polymer generation: **a** multiple properties; **b** polymer structures; **c** natural language. The decoding modules in different generation models vary, such as the discriminator in GANs, the decoder in VAEs, and the reverse diffusion process in diffusion models. These constraints guide the decoding module in a generation model to produce desirable polymers

of 10 barrer. These values, along with other properties, can form a vector input to the neural network. A multi-layer perceptron (MLP) can then map these values into a latent space for processing. Different polymers are built from different monomer structures. For instance, polystyrene is named for its monomer styrene, while polyethylene is named for its monomer ethylene. These polymers have distinct use cases due to their differing chemical structures and properties: polystyrene is ideal for rigid, lightweight, and insulating applications like disposable consumer products, while polyethylene is best for flexible, durable, and chemically resistant applications like packaging. Therefore, using their membership as categorical conditions in the neural network is crucial for generating polymers suited to specific applications.

So, property vectors are often a mix of categorical and continuous values that exist on different scales. For example, gas permeability values typically range from 0 to 10,000 barrer, while categorical labels are often binary (0 or 1). It is ineffective to map them together into the latent space using a single MLP. Therefore, we need to process different dimensions of the vector differently. Categorical values can be one-hot encoded and then embedded into the latent space using neural networks like MLPs. Continuous values require more careful processing, with three practical options: (1) direct embedding, (2) interval-based embeddings, and (3) clustering-based embeddings (Liu et al. 2024).

Direct embedding leverages the neural network's power to learn the embedding space from different numbers. Interval-based embeddings transform continuous value embedding into a multi-class embedding problem by dividing the label space into intervals. Clustering-based embedding groups different values into clusters in the latent space, assigning each value a "soft" one-hot encoding based on softmax. Another linear layer with activation can then produce the final embedding based on cluster assignment. To integrate all property dimensions, we can sum them in a high-dimensional latent space.

### 3.4.1.2 Polymer Structures

In some cases, generated polymers need to contain certain structures that are the same as or similar to known ones. This requirement can be applied in two ways: hard or soft. The hard approach guarantees that the generated polymer must include these specific structures. The soft approach uses the information from these structures without guaranteeing their presence.

For the hard approach, caution is needed when using neural network methods. If the network is not sufficiently trained, it cannot ensure that the input conditions are met. A better approach is to fix part of the graph structure as the desired one and randomly initialize only the remaining structures. The neural network then denoises the randomly initialized structure while keeping the fixed structure unchanged throughout the decoding process.

The soft approach involves using structure constraints without guaranteeing their inclusion. Here, additional graph neural networks (GNNs) can be used to encode the structure and obtain atom representations. These representations can then serve as inputs for the decoding

## 3.4 Constrained Polymer Generation

modules in the generation model. The use of hard or soft approaches depends on whether a particular structure is a strict requirement. For example, researchers may be interested in a specific functional group to ensure the generated polymer has certain properties.

### 3.4.1.3 Natural Languages

Natural language can be used to control polymer generation by converting text data into vectors through an encoder module. Many text encoding neural networks, such as BERT (Devlin et al. 2018) and RoBERTa (Liu et al. 2019), have been developed for this purpose. Among these, SciBERT (Beltagy et al. 2019), which is trained on a large scientific corpus, is particularly promising for understanding descriptions of polymers and molecules. While pretraining such a model requires substantial resources, these text encoder models are often open-sourced and can be directly loaded from platforms like GitHub or Hugging Face. This allows us to implement them to obtain a vector representing the textual requirements.

It is not always necessary to transform all constraints into vectors; this depends on how the constraints are added to the generation models. Constraints can be incorporated either as new loss functions or as additional features to interact with the latent polymer representation during decoding. For the former, transforming constraints into vectors is not required. For the latter approach, different types of constraints are first mapped into high-dimensional latent spaces using projection neural networks, such as an MLP. After incorporating these constraints into the generative models, the model must be trained to generate polymers that adhere to these constraints.

### 3.4.1.4 Adding Constraints as Loss Function

We can use new constraints to construct new loss functions. This approach is typically applicable to both property and polymer structure constraints. Here we show a few examples on how to use constraints as additional loss functions.

To regularize polymer generation, we can add a property predictor, similar to the discriminator module in GANs. This predictor, which can be any neural network discussed in Chap. 2, takes the generated polymers as input and predicts their properties. We can use a multi-task property predictor to handle multiple properties simultaneously or use separate predictors for each task. Alternatively, different relevant properties can be combined into a single predictor. For structural constraints, we can define a reward function that outputs 1 if the desired structure exists in the generated polymer and 0 if it does not. This approach helps ensure that the generated polymers meet specific structural criteria.

Property predictors usually need to be pre-trained for several epochs before being used to construct the loss function for generation constraints. If the property predictor cannot accurately predict the properties of the generated polymers, the calculated loss is meaningless and cannot optimize the generation models effectively. Therefore, pretraining the property predictor is crucial to prevent the generation from diverging.

When using property predictor models for calculating new loss functions and evaluations, it is essential not to use the same model for both purposes. The property predictor should be trained separately from the test set to avoid data leakage concerns.

Property predictors, typically parameterized by neural networks, are differentiable, allowing backpropagation during loss calculation. However, practical issues may arise: (1) inconsistency between polymer generation output and prediction model input, and (2) the property predictor could be a non-differentiable function. For the first case, the polymer generation output from neural networks usually consists of softmax probabilities for different atom and bond types, while the property predictor input might vary from fingerprint vectors to RDKit tool-extracted features of atoms and bonds. This conversion can disrupt backpropagation. For the second case, useful and computationally efficient metrics like synthetic accessibility are often non-differentiable.

To address the first issue, we can constrain the input of the prediction model during pre-training and apply Gumbel-Softmax sampling for the generation output (Jang et al. 2016), using the reparameterization trick to compute parameter gradients. For the second issue, optimization approaches from reinforcement learning and Bayesian optimization can be used to handle non-differentiable metrics.

When using constraints as a loss function, we iteratively refine the generated polymers by optimizing their structure to meet desired properties during the generation process. In the next section, we will discuss how to use constraints as additional features to directly control generation and produce desirable polymers in a single step.

### 3.4.2 Adding Constraints as Features

To add constraints as additional features, we first need to map different constraints into latent spaces, as illustrated in the previous section. As shown in Fig. 3.4, these constraint vectors can then modulate the polymer hidden states using adaptive normalization techniques or cross-attention mechanisms.

Layer normalization is a popular technique in modern neural networks, such as Transformers. Suppose there are $M$ constraints for each polymer generation. We first average all these vectors into a single vector, termed $\mathbf{c}$. Given a polymer latent matrix $\mathbf{X}$ containing all node representations, layer normalization standardizes it using the formula $\sqrt{\mathrm{Var}[\mathbf{X}] + \epsilon} \cdot (\mathbf{X} - \mathbb{E}[\mathbf{X}]) \cdot \gamma + \beta$, where $\gamma$ and $\beta$ are learnable parameters representing the new mean and variance. We can link the calculation of $\gamma$ and $\beta$ to the constraint vector $\mathbf{c}$. This modifies the layer normalization formula to $\sqrt{\mathrm{Var}[\mathbf{X}] + \epsilon} \cdot (\mathbf{X} - \mathbb{E}[\mathbf{X}]) \cdot \gamma(\mathbf{c}) + \beta(\mathbf{c})$, where $\gamma(\mathbf{c})$ and $\beta(\mathbf{c})$ are neural networks that take the constraints as input and output new mean and variance values. In Diffusion Transformer models, such as OpenAI's video generative model Sora, $\gamma(\mathbf{c})$ and $\beta(\mathbf{c})$ are implemented as linear layers with a SiLU activation function. Initially, their parameters are set to zero, indicating no influence on the latent representation at the beginning of training.

## 3.5 Summary

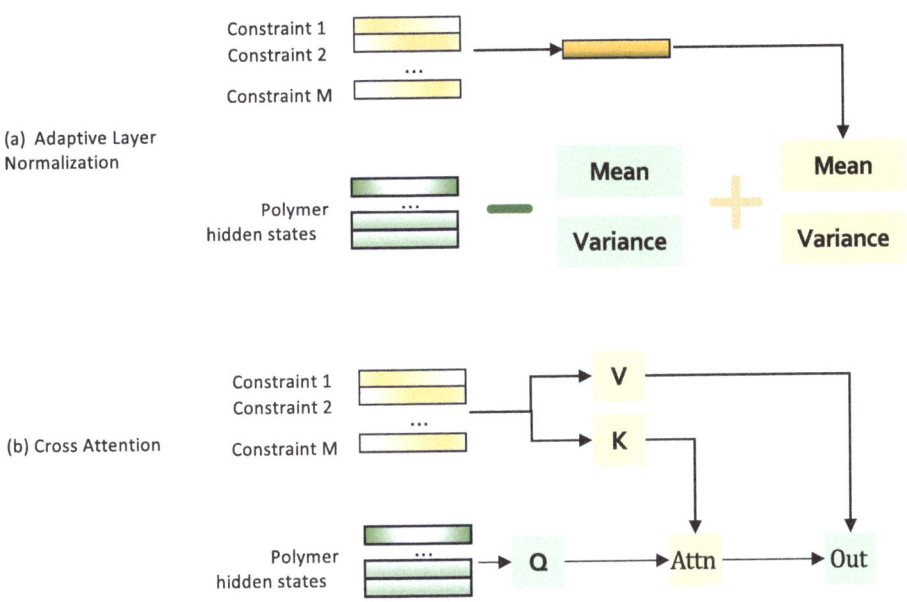

**Fig. 3.4** Different approaches to incorporating constraint features into polymer hidden states during generation include: **a** adaptive layer normalization, and **b** cross-attention mechanism

Alternatively, we can use a cross-attention mechanism. This approach does not require summing all constraint vectors from properties, structures, and natural language into a single vector. Instead, we maintain their individual dimensions and concatenate them into a constraint matrix **C**. In the cross-attention mechanism, the hidden state of the polymers **H**, containing all node representations, serves as the queries in the attention mechanism, while the constraint matrix **C** serves as the keys and values. This allows the polymer hidden states to be updated under the constraints, effectively incorporating the various types of information into the generation process.

## 3.5 Summary

In this chapter, we discussed how to control polymer generation to meet various human requirements, from numerical values to natural language. To handle these constraints, an encoder module is needed–different from the one used in Chapter 2. The constraint encoder takes inputs of categorical, continuous, and text data and outputs latent vectors. These vectors guide the generation model to produce desirable polymers. We explored several neural network architectures for generation, including Generative Adversarial Networks (GANs), Variational Autoencoders (VAEs), and diffusion models, which can be adapted into sequence-based or graph-based generation frameworks. In sequence-based polymer generation, the input sequence is tokenized into units. In graph-based generation, the units could be atoms, bonds, or subgraphs. To incorporate constraints into a specific generative neural network, they can be used either as constraints or features.

## References

J. Austin, D. D. Johnson, J. Ho, D. Tarlow, and R. Van Den Berg. Structured denoising diffusion models in discrete state-spaces. *Advances in Neural Information Processing Systems*, 34:17981–17993, 2021.

J. W. Barnett, C. R. Bilchak, Y. Wang, B. C. Benicewicz, L. A. Murdock, T. Bereau, and S. K. Kumar. Designing exceptional gas-separation polymer membranes using machine learning. *Science advances*, 6(20):eaaz4301, 2020.

I. Beltagy, K. Lo, and A. Cohan. Scibert: A pretrained language model for scientific text. *arXiv preprint* arXiv:1903.10676, 2019.

G. W. Bemis and M. A. Murcko. The properties of known drugs. 1. molecular frameworks. *Journal of medicinal chemistry*, 39(15):2887–2893, 1996.

H. Chen, O. Engkvist, Y. Wang, M. Olivecrona, and T. Blaschke. The rise of deep learning in drug discovery. *Drug discovery today*, 23(6):1241–1250, 2018.

J. Degen, C. Wegscheid-Gerlach, A. Zaliani, and M. Rarey. On the art of compiling and using'drug-like'chemical fragment spaces. *ChemMedChem*, 3(10):1503, 2008.

J. Devlin, M.-W. Chang, K. Lee, and K. Toutanova. Bert: Pre-training of deep bidirectional transformers for language understanding. *arXiv preprint* arXiv:1810.04805, 2018.

W. Gao, T. Fu, J. Sun, and C. Coley. Sample efficiency matters: a benchmark for practical molecular optimization. *Advances in neural information processing systems*, 35:21342–21357, 2022.

I. Goodfellow, J. Pouget-Abadie, M. Mirza, B. Xu, D. Warde-Farley, S. Ozair, A. Courville, and Y. Bengio. Generative adversarial networks. *Communications of the ACM*, 63(11):139–144, 2020.

J. Ho, A. Jain, and P. Abbeel. Denoising diffusion probabilistic models. *Advances in neural information processing systems*, 33:6840–6851, 2020.

E. Jang, S. Gu, and B. Poole. Categorical reparameterization with gumbel-softmax. *arXiv preprint* arXiv:1611.01144, 2016.

W. Jin, R. Barzilay, and T. Jaakkola. Junction tree variational autoencoder for molecular graph generation. In *International conference on machine learning*, pages 2323–2332. PMLR, 2018.

W. Jin, R. Barzilay, and T. Jaakkola. Hierarchical generation of molecular graphs using structural motifs. In *International conference on machine learning*, pages 4839–4848. PMLR, 2020.

J. Jo, S. Lee, and S. J. Hwang. Score-based generative modeling of graphs via the system of stochastic differential equations. In *International Conference on Machine Learning*, volume 162, pages 10362–10383. PMLR, 2022.

D. P. Kingma. Auto-encoding variational bayes. *arXiv preprint* arXiv:1312.6114, 2013.

G. Liu, J. Xu, T. Luo, and M. Jiang. Graph diffusion transformers for multi-conditional molecular generation. In *The Thirty-eighth Annual Conference on Neural Information Processing Systems*, 2024.

Y. Liu, M. Ott, N. Goyal, J. Du, M. Joshi, D. Chen, O. Levy, M. Lewis, L. Zettlemoyer, and V. Stoyanov. Roberta: A robustly optimized bert pretraining approach. *arXiv preprint* arXiv:1907.11692, 2019.

S. Otsuka, I. Kuwajima, J. Hosoya, Y. Xu, and M. Yamazaki. Polyinfo: Polymer database for polymeric materials design. In *2011 International Conference on Emerging Intelligent Data and Web Technologies*, pages 22–29. IEEE, 2011.

# References

A. Ramesh, M. Pavlov, G. Goh, S. Gray, C. Voss, A. Radford, M. Chen, and I. Sutskever. Zero-shot text-to-image generation. In *International conference on machine learning*, pages 8821–8831. Pmlr, 2021.

R. Rombach, A. Blattmann, D. Lorenz, P. Esser, and B. Ommer. High-resolution image synthesis with latent diffusion models. In *Proceedings of the IEEE/CVF conference on computer vision and pattern recognition*, pages 10684–10695, 2022.

T. Salimans, I. Goodfellow, W. Zaremba, V. Cheung, A. Radford, and X. Chen. Improved techniques for training gans. *Advances in neural information processing systems*, 29, 2016.

A. Thornton, L. Robeson, B. Freeman, and D. Uhlmann. Polymer gas separation membrane database, 2012. URL https://research.csiro.au/virtualscreening/membrane-database-polymer-gas-separation-membranes/.

C. Vignac, I. Krawczuk, A. Siraudin, B. Wang, V. Cevher, and P. Frossard. Digress: Discrete denoising diffusion for graph generation. *arXiv preprint* arXiv:2209.14734, 2022.

# Interpretable Learning: Graph Rationalization with Environment-Based Augmentation

## 4.1 Introduction

Graph property prediction has attracted attention in different research fields like chemoinformatics and bioinformatics where small molecules are represented as labelled graphs of atoms (Hu et al. 2020; Zhou et al. 2020; Guo et al. 2021). Besides, materials informatics for *polymers* has emerged in recent years from property prediction to inverse design (Kim et al. 2018; Chen et al. 2021). Polymers are materials consisting of macromolecules, composed of many repeating units. They are ubiquitous in applications ranging from plastic cups and electronics to aerospace structures. New engineering and environmental challenges demand that polymers possess unconventional properties such as high-temperature stability, excellent thermal conductivity, and biodegradability (Ma et al. 2019; Wei et al. 2021). It's important to integrate data science and machine learning into polymer informatics on the tasks of graph classification and regression.

To automate feature extraction from graph data, graph neural network (GNN) models learn node representations through nonlinear functions and layers that aggregate information from node neighborhood (Kipf and Welling 2017; Veličković et al. 2018; Hamilton et al. 2017; Zhang et al. 2020; Wu et al. 2020). Graph pooling is a central component of the GNN architecture that learns a cluster assignment for nodes and passes cluster nodes and their representations to the next layer (Ying et al. 2018; Lee et al. 2019). The final layer returns the representations of entire graphs. Despite the advances of various GNN models, the limitation of data size makes them easily fall into *over-fitting and poor generalizability*. For example, the number of graphs in molecule benchmark datasets is usually ranging from 1,000 to 10,000 and the size of polymer datasets is even smaller (e.g., ∼600) (Ma and Luo 2020).

Rationalization techniques have been designed to solve the above problem in vision and language data, where the rationale is defined as a subset of input features that best explains or supports the prediction by machine learning models (Chang et al. 2020; Arjovsky et al. 2019; Rosenfeld et al. 2021). However, graph rationalization has not been extensively studied, which aims at identifying representative subgraph structures for accurate and interpretable graph property prediction. Related work mainly focused on advancing graph pooling methods, but cluster assignment could not reflect the most essential part that led to accurate prediction (Mesquita et al. 2020; Gao and Ji 2021). A very recent technique named DIR (Wu et al. 2022) employed two GNN modules to discover invariant graph rationales: one module separates each input graph into a rationale subgraph and an environment subgraph; the other is a graph property predictor based on the rationale subgraph. As shown at the top in Fig. 4.1, given graph $g_i$, the separator $f_{sep}$ identifies rationale $g_i^{(r)}$, and the predictor $f_{pred}$ gives label $\hat{y}_i^{(r)}$ based on the rationale. DIR conducted interventions on training distribution to improve the invariance. Unfortunately, when the data size was small, $f_{sep}$ could hardly find good rationales, as reported in later experiments.

In this chapter, we introduce the technique to enhance graph rationalization by graph data augmentations. Existing augmentation methods were mainly heuristic modification of graph structure, which could not directly support the identification of graph rationales (Rong et al. 2019; Wang et al. 2020a; Wang et al. 2020b; Zhao et al. 2021). We present two augmentation methods based on *environment subgraphs* that are the remaining parts in the graph after rationale identification. First, rationales are used to train the property predictor, which can be considered as graph examples augmented by *environment removal*. Second,

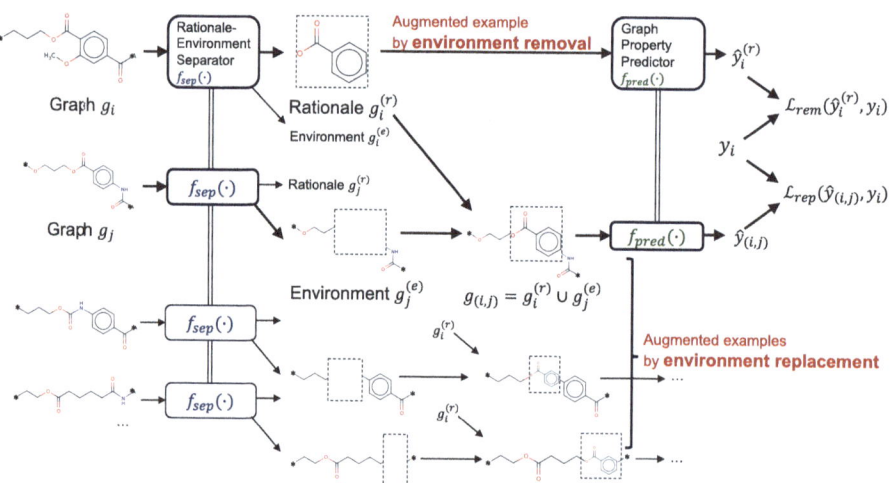

**Fig. 4.1** Graph rationalization identifies a rationale subgraph that best explains or supports the prediction of graph property. This framework makes the first attempt to improve graph rationalization by graph data augmentations with *environment subgraphs* which are the remaining parts after rationale identification. It has new augmentation operations, designs and develops a novel graph rationalization framework, and conducts experiments on a large set of molecule and polymer data

## 4.1 Introduction

we replace the environment of input graph with the environment of another graph in the batch: to generate an augmented example: this augmentation method is called *environment replacement*. The idea is that the rationale can be accurately identified and/or separated from the input graph when the augmented examples are expected to have the same label of the input graph example.

Figure 4.1 presents the idea of generating virtual data for small datasets via data augmentations. Suppose we have rationale $g_i^{(r)}$ separated from input graph $g_i$. We use the same GNN-based separator to find environment subgraph $g_j^{(e)}$ from another graph $g_j$ in the batch. The example augmented by environment replacement is denoted by $g_{(i,j)} = g_i^{(r)} \cup g_j^{(e)}$. The model is trained on this example to predict label $\hat{y}_{i,j}$ to be the same as $y_i$ that is the observed label of $g_i$. We compute two losses on the augmented examples, $\mathcal{L}_{rem}$ and $\mathcal{L}_{rep}$ ("rem" for removal and "rep" for replacement), and jointly optimize $f_{sep}$ and $f_{pred}$ by their combination.

The key challenge in the idea implementation is the high computational complexity of decoding for *explicit graph forms* of rationales, environment subgraphs, and augmented examples, as well as encoding them for representation learning and property prediction. Moreover, it is scientifically and technically difficult to explicitly combine rationale $g_i^{(r)}$ and environment $g_j^{(e)}$ from different graphs, as shown in the three augmented examples $g_{(i,j)}$ in Fig. 4.1. To address these challenges, we hypothesize that the *contextualized representations of nodes* play a significant role in rationales, environment subgraphs, and augmented graphs. Thus, we create the representations of all these objects from *one latent space*.

In this chapter, we introduce an efficient framework of Graph Rationalization enhanced by Environment-based Augmentations (GREA). It performs rationale-environment separation and representation learning on the real and augmented examples in one latent space to avoid the high complexity of explicit subgraph decoding and encoding. Figure 4.2 presents the

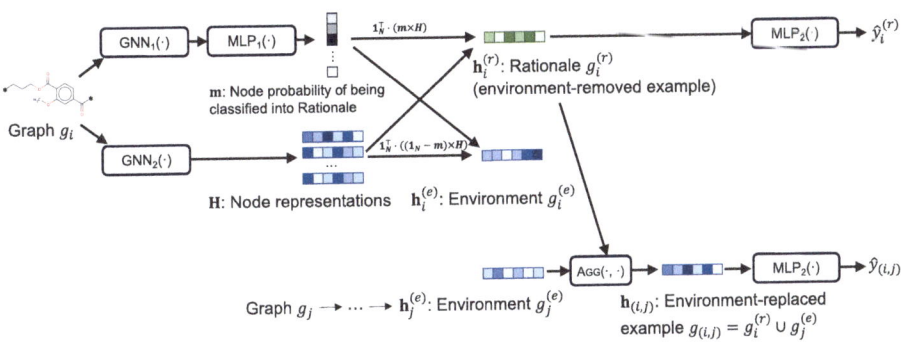

**Fig. 4.2** The architecture of the graph rationalization framework: It performs the creation and representation learning of environment-based augmented examples in a *latent space*, instead of decoding every example into a graph form and running a GNN encoder on it. This design aligns graph representation spaces and avoids high computational complexity

architecture of GREA with a few steps. First, it employs GNN$_1$ and MLP$_1$ models to infer the probability of nodes being classified into rationale subgraph **m**. Second, it employs GNN$_2$ to create contextualized node representations **H**. Then, it *directly* creates the representation vectors of rationales, environment subgraphs and environment-replaced examples, denoted by $\mathbf{h}_i^{(r)}$, $\mathbf{h}_i^{(e)}$, and $\mathbf{h}_{(i,j)}$, respectively. Note that DIR (Wu et al. 2022) used a GNN to generate a matrix of masks that indicate the importance of edges and then select the top-$K$ edges with the highest masks to construct the rationale. Then it had to run GNNs on all the explicit graph objects. Instead, GREA uses **m** and **H** to compute the representation vectors of the artificial graphs.

We present experiments on seven molecule and four polymer datasets. Results demonstrate the advantages of GREA over baselines. For example, it significantly reduces the prediction error on oxygen permeability of polymer membrane with only 595 training examples. The oxygen permeability defines how easily oxygen passes through a particular material. Accurate prediction will speed up material discovery for healthcare and energy utilization.

The main content of this chapter is summarized below:

- We focus on graph rationale identification using data augmentations, including environment replacement, for accurate and interpretable property prediction.
- We introduce a framework that performs rationale-environment separation and representation learning on real and augmented examples in one latent space;
- We conduct extensive experiments on more than ten molecule and polymer datasets to demonstrate the effectiveness and efficiency of the GREA framework.

## 4.2 Problem Definition

### 4.2.1 Graph Property Prediction

Let $g = (\mathcal{V}, \mathcal{E})$ be a graph of $N$ nodes and $M$ edges, where $\mathcal{V}$ is the set of nodes (e.g., atoms) and $\mathcal{E} \subseteq \mathcal{V} \times \mathcal{V}$ is the set of edges (e.g., bonds between atoms). We use $y \in \mathcal{Y}$ to denote the graph-level property of $g$, where $\mathcal{Y}$ is the value space. It can have a categorical or numerical value, corresponding to the task of classification or regression, respectively.

A graph property predictor $f_{pred}$ takes a graph $g$ as input and predicts its label $\hat{y}$. Specifically, a GNN-based predictor employs a GNN encoder to generate node representations **H** from $g$:

$$\mathbf{H} = \begin{bmatrix} - \\ \mathbf{h}_v \\ - \end{bmatrix}_{v \in \mathcal{V}} = \text{GNN}(g) \in \mathbb{R}^{N \times d}, \tag{4.1}$$

## 4.2 Problem Definition

where $\mathbf{h}_v \in \mathbb{R}^d$ is the representation vector of node $v$ in graph $g$. There are many choices for the GNN encoder $\text{GNN}(\cdot)$, such as Graph Convolutional Network (GCN) (Kipf and Welling 2017) and Graph Isomorphism Network (GIN) (Xu et al. 2019).

Once the node representations are ready, a multilayer perceptron (MLP) can project them into a one-dimensional space to obtain a scalar for each node:

$$m_v = \text{MLP}(\mathbf{h}_v). \tag{4.2}$$

As we are more interested in graph-level classification or regression, we first use a readout operator (e.g., average pooling) to get the graph representation $\mathbf{h}$:

$$\mathbf{h} = \text{READOUT}(\mathbf{H}) \in \mathbb{R}^d, \tag{4.3}$$

and then we apply a MLP to project it to a graph label:

$$\hat{y} = \text{MLP}(\mathbf{h}) \in \mathcal{Y}. \tag{4.4}$$

### 4.2.2 Graph Rationalization

Following the existing literature on graph rationalization (Ying et al. 2018; Lee et al. 2019; Gao and Ji 2021; Fan et al. 2021; Wu et al. 2022) and GNN explanation (Ying et al. 2019), we use rationale $g^{(r)} = (\mathcal{V}^{(r)}, \mathcal{E}^{(r)})$ to indicate the causal subgraph of the property $y$, where $g^{(r)}$ is a subgraph of $g$ such that $\mathcal{V}^{(r)} \subseteq \mathcal{V}$ and $\mathcal{E}^{(r)} \subseteq \mathcal{E}$. We use $g^{(e)}$ to denote the environment subgraph, which is the complementary subgraph of $g^{(r)}$ in $g$. In contrast with the rationale subgraph $g^{(r)}$, the environment subgraph $g^{(e)}$ corresponds to the non-causal part of the graph data, which has no causal relationship or only spurious correlation with the target graph property (Chang et al. 2020; Wu et al. 2022).

Let $f_{sep}$ be a GNN-based graph rationalization model that splits an input graph $g$ into a rationale subgraph $g^{(r)}$ and an environment subgraph $g^{(e)}$. Existing graph rationalization methods used only the rationale subgraph as input for property prediction (Lee et al. 2019; Ying et al. 2018; Gao and Ji 2021; Wu et al. 2022):

$$\hat{y} = \hat{y}^{(r)} = f_{pred}(g^{(r)}), \tag{4.5}$$

where $f_{pred}(\cdot) = \text{MLP}(\text{READOUT}(\text{GNN}(\cdot)))$ and $\hat{y}^{(r)}$ denotes the predicted property of the rationale subgraph $g^{(r)}$.

Unfortunately, when suffering from lack of training examples, these methods chose to discard environment subgraphs at the training stage. The environment subgraphs can provide natural noise through data augmentation to improve graph rationalization. In the next section, we present a novel framework to implement this idea.

## 4.3 Interpretable Graph Neural Networks: GREA

We introduce the graph rationalization framework called GREA. The key idea is to augment the rationale subgraph by removing its own environment subgraph and/or combining it with different environment subgraphs. Figure 4.2 shows the overall architecture of GREA:

- $GNN_1$ and $MLP_1$ separate input graph $g$ into rationale subgraph $g^{(r)}$ and environment subgraph $g^{(e)}$: see Eq. (4.1–4.2);
- $GNN_2$ generates node representations $\mathbf{H}$: see Eq. (4.1);
- the rationale subgraph's representation $\mathbf{h}_i^{(r)}$ is then combined with different environment subgraph's representations $\mathbf{h}_j^{(e)}$ for the augmented graph's representations $\mathbf{h}_{(i,j)}$;
- both $\mathbf{h}_i^{(r)}$ and $\mathbf{h}_{(i,j)}$ are fed into $MLP_2$ for the prediction of $y_i$ during training stage: see Eq. (4.4).

### 4.3.1 Rationale-Environment Separation

To separate input graph $g$ into rationale subgraph $g^{(r)}$ and environment subgraph $g^{(e)}$, the rationale-environment separator consists of two components: a GNN encoder ($GNN_1$) that generates latent node representations and a MLP decoder ($MLP_1$) that maps the node representations to a mask vector $\mathbf{m} \in (0, 1)^N$ on the nodes in the set $\mathcal{V}$. $m_v = Pr(v \in \mathcal{V}^{(r)})$ is the node-level mask that indicates the probability of node $v \in \mathcal{V}$ being classified into the rationale subgraph. The mask can be on either a node or an edge (Wu et al. 2022). We choose to learn masks on the nodes to avoid the computational complexity of edge selection. Hence, $\mathbf{m}$ can be calculated as

$$\mathbf{m} = \sigma(MLP_1(GNN_1(g))), \tag{4.6}$$

where $\sigma$ denotes the sigmoid function. Based on $\mathbf{m}$, we have $(\mathbf{1}_N - \mathbf{m})$ that indicates the probability of nodes being classified into the environment subgraph. $GNN_1$ and $MLP_1$ make up the GNN-based graph rationalization model $f_{sep}$.

GREA uses another GNN encoder to generate contextualized node representations $\mathbf{H}$: $\mathbf{H} = GNN_2(g)$. With $\mathbf{m}$ and $\mathbf{H}$, the rationale subgraph and environment subgraph can be easily separated in the *latent space*. Using sum pooling, we have

$$\mathbf{h}^{(r)} = \mathbf{1}_N^\top \cdot (\mathbf{m} \times \mathbf{H}), \tag{4.7}$$

$$\mathbf{h}^{(e)} = \mathbf{1}_N^\top \cdot ((\mathbf{1}_N - \mathbf{m}) \times \mathbf{H}), \tag{4.8}$$

where $\mathbf{1}_N$ denotes the $N$-size column vector with all entries as 1, and $\mathbf{h}^{(r)}, \mathbf{h}^{(e)} \in \mathbb{R}^d$ are the representation vectors of graph $g^{(r)}$ and $g^{(e)}$, respectively.

## 4.3.2 Environment-Based Augmentations

Suppose $g_1, g_2, \ldots, g_B$ are the input graphs in one batch for training, where $B$ is known as batch size. The rationale-environment separator has generated the graph representations of rationale and environment subgraphs for each graph $g_i$. That is, we have $\{(\mathbf{h}_1^{(r)}, \mathbf{h}_1^{(e)}), (\mathbf{h}_2^{(r)}, \mathbf{h}_2^{(e)}), \ldots, (\mathbf{h}_B^{(r)}, \mathbf{h}_B^{(e)})\}$. We design environment-based augmentations in the latent space of graph representations.

**Environment Removal Augmentation.** As graph rationalization aims to find the rationale subgraph which is regarded as the causal factor of graph property, the rationale itself should be good for property prediction. As in the graph pooling methods (Lee et al. 2019; Gao and Ji 2021) and the graph rationalization as defined in Eq. (4.5), the environment removal augmentation uses the rationale subgraph only for training the graph property predictor. That is, given the rationale subgraph representation $\mathbf{h}_i^{(r)}$ of graph $g_i$, the predicted label is

$$\hat{y}_i^{(r)} = \text{MLP}_2\left(\mathbf{h}_i^{(r)}\right). \tag{4.9}$$

**Environment Replacement Augmentation.** The environment subgraphs can be viewed as natural noises on the rationale subgraphs. Hence, in order to enhance the model's robustness against the noise signal brought by the environment subgraphs, for each graph $g_i$, we combine its rationale subgraph $g_i^{(r)}$ not only with its own environment subgraph $g_i^{(e)}$, but also with all other environment subgraphs $g_j^{(e)}$, $j \in \{1, 2, \ldots, B\} \setminus \{i\}$ in the batch. By replacing the environment subgraph with other environment subgraphs in the batch, the environment replacement augmentation generates $B - 1$ augmented data samples for each graph during training. As the environment replacement happens on the latent space, an aggregation function $\text{AGG}(\cdot, \cdot)$ is used to combine the rationale subgraph representation $\mathbf{h}_i^{(r)}$ and environment subgraph representation $\mathbf{h}_j^{(e)}$. The aggregation function can be any combining/pooling functions such as concatenation, sum pooling, and max pooling. Taking the element-wise sum pooling as an example, the graph representation $\mathbf{h}_{(i,j)}$ of a combined graph of rationale subgraph $g_i^{(r)}$ and environment subgraph $g_j^{(e)}$ can be calculated as below:

$$\mathbf{h}_{(i,j)} = \text{AGG}\left(\mathbf{h}_i^{(r)}, \mathbf{h}_j^{(e)}\right) = \mathbf{h}_i^{(r)} + \mathbf{h}_j^{(e)}. \tag{4.10}$$

For the graph representations $\mathbf{h}_{(i,j)}$ generated by the environment replacement augmentation, the MLP property predictor is trained to predict $y_i$. That is,

$$\hat{y}_{(i,j)} = \text{MLP}_2\left(\mathbf{h}_{(i,j)}\right). \tag{4.11}$$

### 4.3.3 Optimization

During training, the type of loss function on the observed graph property ($y_i$) and predicted labels ($\hat{y}_i^{(r)}$ and $\hat{y}_{(i,j)}$) depends on the type of the property label. For example, when the graph property $y$ has binary values in the binary classification task, we use the standard binary cross-entropy loss. When the graph property $y$ has real values in the graph regression task, we use the mean squared error (MSE) loss. Without loss of generality, suppose we focus on the binary classification task. Given a batch of $B$ graphs $g_1, g_2, \ldots, g_B$, the loss functions for each graph example $g_i$ and its label $y_i$ are defined as

$$\mathcal{L}_{rem} = y_i \cdot \log \hat{y}_i^{(r)} + (1 - y_i) \cdot \log\left(1 - \hat{y}_i^{(r)}\right), \tag{4.12}$$

$$\mathcal{L}_{rep} = \frac{1}{B} \sum_{j=1}^{B} \left( y_i \cdot \log \hat{y}_{(i,j)} + (1 - y_i) \cdot \log(1 - \hat{y}_{(i,j)}) \right), \tag{4.13}$$

where $\mathcal{L}_{rem}$ is the loss for the examples created by environment removal augmentation, and $\mathcal{L}_{rep}$ is the loss for the examples created by the environment replacement augmentation.

The graph representations generated by both environment removal augmentation and environment replacement augmentation (i.e., $\mathbf{h}_i^{(r)}$ and $\mathbf{h}_{(i,j)}$) are fed into the same property predictor MLP$_2$. The GNN-based property predictor $f_{pred}$ includes MLP$_2$ and GNN$_2$ that generates the contextualized node representation $\mathbf{H}$.

Moreover, the following regularization term is used to control the size of the selected rationale subgraph:

$$\mathcal{L}_{reg} = \left( \frac{\mathbf{1}_N^\top \mathbf{m}}{N} - \gamma \right) + \left( \frac{\sum_{k:\mathbf{m}_k>0} 1}{N} - \gamma \right), \tag{4.14}$$

where $\gamma \in [0, 1]$ controls the expected size of the rationale subgraph $g^{(r)}$. The first term penalizes the number of nodes in the rationale when it deviates from the expectations. The second term encourages an uneven distribution for $\mathbf{m}$.

We use the alternate training schema (Chang et al. 2020) to train GREA. That is, we iteratively train $f_{sep}$ (GNN$_1$ and MLP$_1$) and $f_{pred}$ (GNN$_2$ and MLP$_2$) for a fixed number of epochs $T_{sep}$ and $T_{pred}$, respectively. The loss functions for training GREA are

$$\mathcal{L}_{pred} = \mathcal{L}_{rem} + \alpha \cdot \mathcal{L}_{rep}, \tag{4.15}$$

$$\mathcal{L}_{sep} = \mathcal{L}_{rem} + \alpha \cdot \mathcal{L}_{rep} + \beta \cdot \mathcal{L}_{reg}, \tag{4.16}$$

where $\mathcal{L}_{pred}$ in Eq. (4.15) and $\mathcal{L}_{sep}$ in Eq. (4.16) are used to train $f_{sep}$ (GNN$_1$ and MLP$_1$) and $f_{pred}$ (GNN$_2$ and MLP$_2$), respectively. $\alpha$ and $\beta$ are hyperparameters that control the weights of $\mathcal{L}_{rep}$ and $\mathcal{L}_{reg}$, respectively. During inference, $\hat{y}_i^{(r)}$ is used as the final predicted property of input graph $g_i$.

## 4.4 Experiments

We conduct experiments to answer the following questions:
- **(RQ1)** Effectiveness: Does the GREA make more accurate prediction on molecule and polymer properties than existing graph classification/regression methods?
- **(RQ2)** Ablation study: Do the environment-based augmentations make positive effect on the performance?
- **(RQ3)** Case study: Based on domain expertise, are the polymer rationale examples identified by GREA representative?
- **(RQ4)** Efficiency: Does the *latent space-based design* for augmentations perform faster than explicit graph decoding and encoding? Can we empirically analyze the complexity?
- **RQ5)** Sensitivity analysis: Is the performance of GREA sensitive to hyperparameters such as $\alpha$, $\beta$, and AGG($\cdot$)?

### 4.4.1 Experimental Settings

**Datasets.** We conduct experiments on **four** polymer datasets and **seven** molecule datasets. The statistics of the datasets are given in Table 4.1, such as number of graphs and average size of graphs.

- Polymer datasets: The four datasets GlassTemp, MeltingTemp, PolyDensity, and $O_2$Perm are used to predict different properties of polymers such as *glass transition temperature* (°C), *polymer density* g/cm$^3$, *melting temperature* (°C), and *oxygen permeability* (Barrer). GlassTemp, MeltingTemp, and PolyDensity are collected from PolyInfo, which is the largest web-based polymer database (Otsuka et al. 2011). The $O_2$Perm dataset is created from the Membrane Society of Australasia portal, consisting of a variety of gas permeability data (Thornton et al. 2012). However, the limited size (i.e., 595 polymers) brings great challenges to rationale identification and property prediction. Since a polymer is built from repeated monomer units, researchers use monomers as polymer graphs to predict properties. Different from molecular graphs, the monomer graphs have two special nodes (see "∗" in the molecular structures in Fig. 4.1), indicating the polymerization points of monomers (Ma and Luo 2020). For all the polymer datasets, we randomly split by 60%/10%/30% for training, validation, and test.
- Molecule datasets: Besides polymer datasets, we use seven molecule datasets from the graph property prediction task on Open Graph Benchmark or known as OGBG. They were originally collected by MoleculeNet (Wu et al. 2018) and used to predict the properties of molecules, including (1) inhibition to HIV virus replication in ogbg-HIV, (2) toxicological properties of 617 types in ogbg-ToxCast, (3) toxicity measurements such as nuclear receptors and stress response in ogbg-Tox21, (4) blood–brain barrier permeability

**Table 4.1** Statistics of eleven datasets for graph property prediction: the four top rows are polymer datasets

| Dataset | # Graphs | Avg./max # Nodes | Avg./max # Edges |
|---|---|---|---|
| GlassTemp | 7,174 | 36.7/166 | 79.3/362 |
| MeltingTemp | 3,651 | 26.9/102 | 55.4/212 |
| PolyDensity | 1,694 | 27.3/93 | 57.6/210 |
| $O_2$Perm | 595 | 37.3/103 | 82.1/234 |
| ogbg-HIV | 41,127 | 25.5/222 | 54.9/502 |
| ogbg-ToxCast | 8,576 | 18.8/124 | 38.5/268 |
| ogbg-Tox21 | 7,831 | 18.6/132 | 38.6/290 |
| ogbg-BBBP | 2,039 | 24.1/132 | 51.9/290 |
| ogbg-BACE | 1,513 | 34.1/97 | 73.7/202 |
| ogbg-ClinTox | 1,477 | 26.2/136 | 55.8/286 |
| ogbg-SIDER | 1,427 | 33.6/492 | 70.7/1010 |

The prediction tasks are graph regression. The seven bottom rows are molecule datasets. Their tasks are graph classification

in ogbg-BBBP, (5) inhibition to human $\beta$-secretase 1 in ogbg-BACE, (6) FDA approval status or failed clinical trial in ogbg-ClinTox, and (7) having drug side effects of 27 system organ classes in ogbg-SIDER. For all molecule datasets, we use the scaffold splitting procedure as OGBG adopted (Hu et al. 2020). It attempts to separate structurally different molecules into different subsets, which provides a more realistic estimate of model performance in experiments (Wu et al. 2018).

**Evaluation Metrics**. On the polymer datasets, we perform the tasks of graph regression. We use the coefficient of determination ($R^2$) and Root Mean Square Error (RMSE) as evaluation metrics according to previous works (Ma and Luo 2020; Hu et al. 2020). On the molecule datasets, we perform the tasks of graph binary classification using the Area under the ROC curve (AUC) as the metric. To evaluate model efficiency, we use the computational time per training batch (in seconds).

**Baseline Methods**. There are three categories of related methods that we can compare GREA with. The first category is *graph pooling* methods that aim at finding (soft) cluster assignment of nodes towards aggregated representations of graph. They are U-NetsPool (Gao and Ji 2021) and SelfAttnPool (Lee et al. 2019). The second category improves the *optimization and generalization* of learned representations. They include StableGNN (Fan et al. 2021), OOD-GNN (Li et al. 2021), and IRM (Arjovsky et al. 2019). The third is DIR for *graph rationale identification* that was developed in a very recent work by Wu et al. (2022). To investigate the effect of *environment replacement augmentation* (denoted by REPAUG as a module that may be used or not in the methods), we implement two method variants:

(1) DIR+REPAUG: We add environment-replaced augmentation to DIR (Wu et al. 2022) to identify rationales, however, it has to explicitly decode and encode the rationales; (2) GREA−REPAUG: We disable the environment replacement augmentation and use only the environment removal augmentation, i.e., rationale subgraphs in GREA. In the experiments, we study two types of GNN models (GCN (Kipf and Welling 2017) and GIN (Xu et al. 2019)) as graph encoders for all the methods.

### 4.4.2 RQ1: Results on Effectiveness

Table 4.2 presents the results on polymer property regression with $R^2$ and RMSE metrics. Table 4.3 presents the results on molecule property classification using AUC. Underlined are for the best baseline(s). The best baseline is OOD-GNN for its elimination of the statistical dependence between property-relevant graph representation and property-irrelevant graph representation. The first graph rationalization method DIR was evaluated on synthetic data (Wu et al. 2022); unfortunately, it performs poorly on real polymer and molecule datasets because it selects edges to create rationale subgraphs and thus loses the original contextual information of atoms in the the rationale representations. Compared to them, GREA with either GCN or GIN consistently achieves the best performance on all the polymer and molecule datasets. On the PolyDensity dataset, GREA with GCN improves $R^2$ over OOD-GNN relatively by +3.91%. On MeltingTemp, GREA with GIN produces $1.56\times R^2$ over DIR.

### 4.4.3 RQ2: Ablation Study on GREA

Tables 4.2 and 4.3 have presented the results of DIR+REPAUG and GREA−REPAUG. DIR+REPAUG is a variant of baseline method DIR by enabling *environment replacement augmentations* for training. GREA−REPAUG is a variant of GREA that disables the replacement augmentations and uses *environment removal* only for training. Clearly, DIR+REPAUG outperforms DIR, showing positive effect of the replacement augmentations. And the performance of GREA−REPAUG is not satisfactory. Environment replacement augmentations are effective for training graph rationalization methods.

### 4.4.4 RQ3: Case Study on Polymer Data

Given test polymer examples in the $O_2$Perm dataset, we visualize and compare the rationale subgraphs that are identified by from DIR (Wu et al. 2022) and GREA in Fig. 4.3. We have three observations.

**Table 4.2** Results on polymer property prediction: GREA consistently achieves the highest $R^2$ and smallest RMSE

| | | GlassTemp | | MeltingTemp | | PolyDensity | | $O_2$Perm | |
|---|---|---|---|---|---|---|---|---|---|
| | | $R^2$ ↑ | RMSE ↓ | $R^2$ ↑ | RMSE ↓ | $R^2$ ↑ | RMSE ↓ | $R^2$ ↑ | RMSE ↓ |
| GCN (Kipf and Welling 2017) as encoder | U-NetsPool (Gao and Ji 2021) | 0.839 ± 0.005 | 44.9 ± 0.7 | 0.685 ± 0.012 | 63.4 ± 1.2 | 0.615 ± 0.053 | 0.100 ± 0.007 | 0.833 ± 0.084 | 865 ± 214 |
| | SelfAttnPool (Lee et al. 2019) | 0.848 ± 0.007 | 43.5 ± 1.0 | 0.709 ± 0.008 | 61.0 ± 0.9 | 0.688 ± 0.019 | 0.090 ± 0.003 | 0.656 ± 0.135 | 1251 ± 266 |
| | StableGNN (Fan et al. 2021) | 0.809 ± 0.013 | 48.8 ± 1.6 | 0.635 ± 0.033 | 70.0 ± 4.5 | 0.667 ± 0.070 | 0.093 ± 0.009 | 0.676 ± 0.127 | 1219 ± 241 |
| | OOD-GNN (Li et al. 2021) | 0.852 ± 0.006 | 43.0 ± 0.9 | 0.714 ± 0.025 | 60.4 ± 2.6 | 0.676 ± 0.010 | 0.092 ± 0.001 | 0.921 ± 0.059 | 576 ± 212 |
| | IRM (Arjovsky et al. 2019) | 0.830 ± 0.008 | 46.1 ± 1.1 | 0.677 ± 0.006 | 64.2 ± 0.6 | 0.690 ± 0.016 | 0.090 ± 0.002 | 0.871 ± 0.043 | 770 ± 141 |
| | DIR (Wu et al. 2022) | 0.697 ± 0.061 | 61.2 ± 6.0 | 0.380 ± 0.214 | 87.8 ± 14. | 0.656 ± 0.036 | 0.094 ± 0.005 | 0.135 ± 0.068 | 2028 ± 80 |
| | DIR+RepAug | 0.800 ± 0.006 | 56.5 ± 3.2 | 0.520 ± 0.101 | 77.8 ± 8.2 | 0.671 ± 0.033 | 0.092 ± 0.005 | 0.915 ± 0.031 | 626 ± 115 |
| | GREA−RepAug | 0.685 ± 0.172 | 60.6 ± 16.5 | 0.679 ± 0.034 | 64.0 ± 3.3 | 0.686 ± 0.007 | 0.090 ± 0.001 | 0.459 ± 0.254 | 1556 ± 395 |
| | GREA (ours) | **0.855 ± 0.003** | **42.6 ± 0.5** | **0.716 ± 0.016** | **60.2 ± 1.6** | **0.717 ± 0.023** | **0.086 ± 0.003** | **0.941 ± 0.018** | **524 ± 91** |

(continued)

## 4.4 Experiments

**Table 4.2** (continued)

| | | GlassTemp | | MeltingTemp | | PolyDensity | | O$_2$Perm | |
|---|---|---|---|---|---|---|---|---|---|
| | | R$^2$ ↑ | RMSE ↓ | R$^2$ ↑ | RMSE ↓ | R$^2$ ↑ | RMSE ↓ | R$^2$ ↑ | RMSE ↓ |
| GIN (Xu et al. 2019) as encoder | U-NETSPOOL (Gao and Ji 2021) | 0.852 ± 0.006 | 42.9 ± 0.9 | 0.703 ± 0.009 | 61.6 ± 0.9 | 0.635 ± 0.029 | 0.097 ± 0.004 | 0.868 ± 0.085 | 753 ± 250 |
| | SELFATTNPOOL (Lee et al. 2019) | 0.848 ± 0.003 | 43.5 ± 0.4 | <u>0.726</u> ± 0.009 | <u>59.2</u> ± 1.0 | 0.654 ± 0.024 | 0.095 ± 0.003 | 0.601 ± 0.267 | 1265 ± 546 |
| | STABLEGNN (Fan et al. 2021) | 0.794 ± 0.007 | 50.8 ± 0.9 | 0.535 ± 0.061 | 76.9 ± 5.0 | 0.642 ± 0.045 | 0.096 ± 0.006 | 0.501 ± 0.266 | 1487 ± 404 |
| | OOD-GNN (Li et al. 2021) | <u>0.862</u> ± 0.007 | <u>41.6</u> ± 1.1 | 0.721 ± 0.006 | 59.7 ± 0.6 | 0.666 ± 0.025 | 0.093 ± 0.003 | <u>0.917</u> ± 0.029 | <u>620</u> ± 109 |
| | IRM (Arjovsky et al. 2019) | 0.842 ± 0.004 | 44.5 ± 0.5 | 0.681 ± 0.008 | 63.8 ± 0.8 | 0.682 ± 0.031 | 0.091 ± 0.004 | 0.890 ± 0.042 | 709 ± 146 |
| | DIR (Wu et al. 2022) | 0.594 ± 0.070 | 71.0 ± 6.0 | 0.287 ± 0.121 | 95.1 ± 7.9 | 0.617 ± 0.045 | 0.099 ± 0.006 | 0.501 ± 0.309 | 1446 ± 537 |
| | DIR+REPAUG | 0.744 ± 0.029 | 56.4 ± 3.2 | 0.542 ± 0.083 | 76.2 ± 7.0 | 0.647 ± 0.058 | 0.095 ± 0.008 | 0.743 ± 0.150 | 1054 ± 338 |
| | GREA−REPAUG | 0.494 ± 0.110 | 79.0 ± 9.3 | 0.660 ± 0.107 | 65.2 ± 9.5 | <u>0.717</u> ± 0.022 | <u>0.086</u> ± 0.003 | 0.400 ± 0.286 | 1623 ± 474 |
| | GREA (ours) | **0.864** ± 0.005 | **41.2** ± 0.8 | **0.736** ± 0.012 | **58.0** ± 1.2 | **0.723** ± 0.030 | **0.085** ± 0.005 | **0.930** ± 0.020 | **569** ± 86 |

Table 4.3 Results on molecule property prediction: GREA consistently achieves the highest AUC (↑)

| | | ogbg-HIV | ogbg-ToxCast | ogbg-Tox21 | IP | ogbg-BACE | ogbg-ClinTox | ogbg-SIDER |
|---|---|---|---|---|---|---|---|---|
| GCN (Kipf and Welling 2017) as encoder | U-NETSPOOL (Gao and Ji 2021) | 0.7527 ± 0.0104 | 0.6507 ± 0.0086 | 0.7492 ± 0.0093 | 0.6709 ± 0.0176 | 0.7757 ± 0.0173 | 0.8450 ± 0.0403 | 0.6181 ± 0.0121 |
| | SELFATTNPOOL (Lee et al. 2019) | 0.7733 ± 0.0187 | 0.6510 ± 0.0076 | 0.7563 ± 0.0080 | 0.6602 ± 0.0220 | 0.7383 ± 0.0541 | 0.8291 ± 0.0791 | 0.5718 ± 0.0219 |
| | STABLEGNN (Fan et al. 2021) | 0.7218 ± 0.0099 | 0.6520 ± 0.0109 | 0.7454 ± 0.0059 | 0.6552 ± 0.0184 | 0.6607 ± 0.0500 | 0.7681 ± 0.0778 | 0.5644 ± 0.0274 |
| | OOD-GNN (Li et al. 2021) | 0.7580 ± 0.0176 | 0.6613 ± 0.0046 | 0.7673 ± 0.0109 | 0.6795 ± 0.0165 | 0.8096 ± 0.0132 | 0.8874 ± 0.0143 | 0.6133 ± 0.0095 |
| | IRM (Arjovsky et al. 2019) | 0.7702 ± 0.0107 | 0.6599 ± 0.0063 | 0.7654 ± 0.0072 | 0.6892 ± 0.0053 | 0.7947 ± 0.0186 | 0.8819 ± 0.0231 | 0.6035 ± 0.0195 |
| | DIR (Wu et al. 2022) | 0.7466 ± 0.0093 | 0.5954 ± 0.0154 | 0.4727 ± 0.0129 | 0.6559 ± 0.0298 | 0.6751 ± 0.0323 | 0.6251 ± 0.0956 | 0.5331 ± 0.0216 |
| | DIR+REPAUG | 0.7494 ± 0.0225 | 0.6632 ± 0.0098 | 0.7437 ± 0.0054 | 0.6630 ± 0.0118 | 0.7677 ± 0.0226 | 0.8606 ± 0.0144 | 0.5934 ± 0.0170 |
| | GREA−REPAUG | 0.7377 ± 0.0210 | 0.6614 ± 0.0048 | 0.7808 ± 0.0061 | 0.6736 ± 0.0077 | 0.7655 ± 0.0529 | 0.8708 ± 0.0514 | 0.6222 ± 0.0166 |
| | GREA (ours) | **0.7794 ± 0.0065** | **0.6662 ± 0.0041** | **0.7822 ± 0.0093** | **0.6986 ± 0.0175** | **0.8191 ± 0.0240** | **0.8961 ± 0.0150** | **0.6316 ± 0.0151** |

(continued)

## 4.4 Experiments

**Table 4.3** (continued)

| | | ogbg-HIV | ogbg-ToxCast | ogbg-Tox21 | IP | ogbg-BACE | ogbg-ClinTox | ogbg-SIDER |
|---|---|---|---|---|---|---|---|---|
| GIN (Xu et al. 2019) as encoder | U-NetsPool (Gao and Ji 2021) | 0.7375 ± 0.0362 | 0.6524 ± 0.0126 | 0.7560 ± 0.0093 | 0.6809 ± 0.0163 | 0.8026 ± 0.0105 | 0.8146 ± 0.0703 | 0.5929 ± 0.0114 |
| | SelfAttnPool (Lee et al. 2019) | 0.7533 ± 0.0247 | 0.6351 ± 0.0137 | 0.7507 ± 0.0110 | 0.6624 ± 0.0167 | 0.7348 ± 0.0194 | 0.7912 ± 0.0995 | 0.5702 ± 0.0137 |
| | StableGNN (Fan et al. 2021) | 0.7218 ± 0.0078 | 0.6485 ± 0.0025 | 0.7381 ± 0.0123 | 0.6695 ± 0.0120 | 0.7229 ± 0.0122 | 0.8559 ± 0.0224 | 0.5593 ± 0.0172 |
| | OOD-GNN (Li et al. 2021) | 0.7799 ± 0.0078 | 0.6697 ± 0.0051 | 0.7646 ± 0.0038 | 0.6710 ± 0.0188 | 0.7800 ± 0.0228 | 0.8416 ± 0.0496 | 0.5916 ± 0.0169 |
| | IRM (Arjovsky et al. 2019) | 0.7817 ± 0.0120 | 0.6641 ± 0.0065 | 0.7542 ± 0.0084 | 0.6835 ± 0.0071 | 0.7977 ± 0.0208 | 0.8485 ± 0.0215 | 0.5778 ± 0.0206 |
| | DIR (Wu et al. 2022) | 0.7533 ± 0.0117 | 0.5927 ± 0.0097 | 0.5078 ± 0.0313 | 0.5843 ± 0.0443 | 0.6115 ± 0.0587 | 0.6911 ± 0.0810 | 0.5406 ± 0.0127 |
| | DIR+RepAug | 0.7725 ± 0.0249 | 0.6454 ± 0.0061 | 0.7453 ± 0.0080 | 0.6813 ± 0.0203 | 0.7590 ± 0.0642 | 0.8561 ± 0.0159 | 0.5730 ± 0.0115 |
| | GREA–RepAug | 0.7770 ± 0.0178 | 0.6681 ± 0.0066 | 0.7690 ± 0.0117 | 0.6737 ± 0.0235 | 0.7997 ± 0.0380 | 0.8574 ± 0.0442 | 0.5988 ± 0.0169 |
| | GREA (ours) | **0.7932 ± 0.0092** | **0.6750 ± 0.0067** | **0.7723 ± 0.0119** | **0.6970 ± 0.0128** | **0.8237 ± 0.0237** | **0.8789 ± 0.0368** | **0.6014 ± 0.0204** |

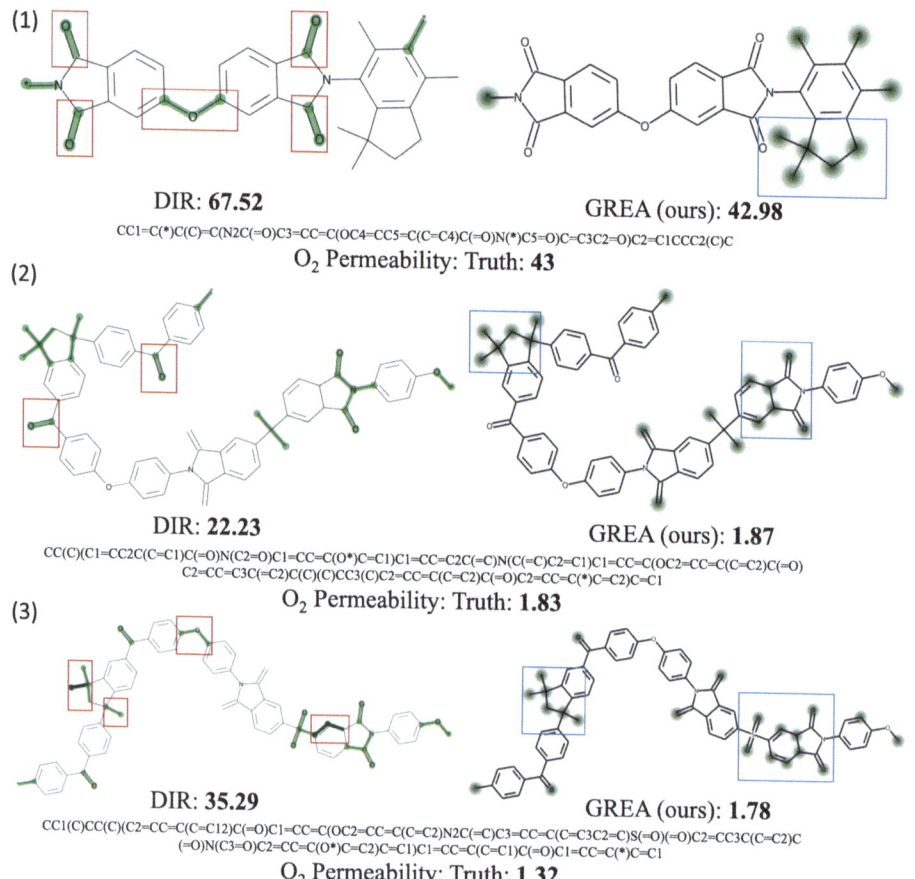

**Fig. 4.3** Three polymer examples in $O_2$Perm test set to compare graph rationales and property predictions by DIR (Wu et al. 2022) and GREA. DIR selects *edges* to decode rationale subgraphs. GREA estimates the probability of *nodes* being classified into rationales in latent space. The red boxes indicate incoherent edges that DIR selects. The blue boxes indicate coherent node sets that contribute to accurate predictions on oxygen permeability of polymer membrane

First, the rationales identified by GREA have more *coherent structures of atom nodes* than those identified by DIR. The red boxes show that quite a few edges in the rationales by DIR are far separated. This is because DIR explicitly decodes the subgraphs by selecting edges. GREA estimates the probability of *nodes* being included in the rationales and uses the *contextualized representations* of atoms in the input graphs to create the representations of rationales. So the rationales have coherent structures of nodes.

Second, the rationales from GREA are *more interpretable and beneficial* than the ones from DIR, based on domain expertise in polymer science. Take a look at the first polymer example in Fig. 4.3. The rationale from GREA includes non-aromatic rings and methyl

## 4.4 Experiments

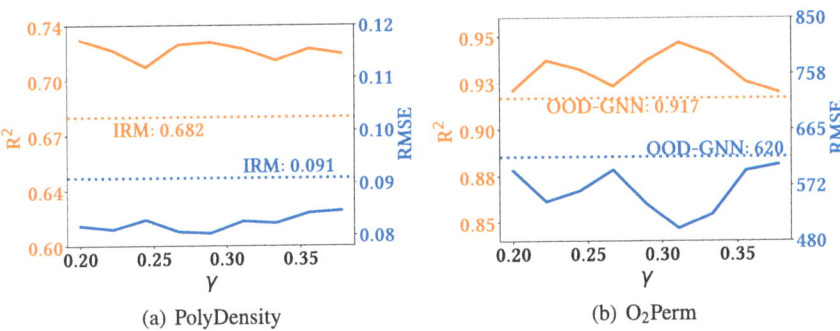

**Fig. 4.4** On two polymer datasets, the performance of GREA is *not* sensitive to rationale size $\gamma$ across a wide tuning range

groups. The former group allows larger free volume elements and lower densities (i.e., enlarge microporousity) in the polymer's repeating units, which positively contributes to the gas permeability (Sanders et al. 2013; Yang et al. 2021). The latter group is hydrophobic and contributes to steric frustration between polymer chains (Yang et al. 2021), inducing a positive correlation to the permeability. On the other hand, the rationale from DIR would make property predictor overestimate the oxygen permeability, because it suggests that the double-bonded oxygens, ethers, and nitrogen atoms are positively correlated with the property. However, it conflicts with observations and conclusions from chemical experiments in previous literature (Yang et al. 2021) where researchers argue that the double-bonded oxygens, ethers, and nitrogen atoms are negatively correlated with gas permeability. For the second and third examples, DIR also predicts through double-bonded oxygens, ethers, and nitrogen atoms, and it overestimates the permeability. GREA realizes and employs the true relationship between the functional groups and property and successfully suppresses the representations of non-aromatic rings and methyl groups in the prediction. GREA intrinsically discovers correct relationships between rationale subgraphs and the property.

Third, the rationales from GREA are *commonly observed across different polymers*. We expect rationales to have universal indication on the polymer properties. The rationales identified in the second and third examples both have the fused heterocyclic rings (at the right end of the monomers and highlighted by blue boxes).

### 4.4.5 RQ4: Results on Efficiency

We conduct efficiency analysis using the ogbg-HIV dataset without losing the generality. Results are presented in Fig. 4.5. When batch size increases, in other words, when a batch has more and more graphs, the time cost per batch of DIR increases significantly; GREA spends much less time than DIR. Empirically we show that GREA is more efficient than DIR. This is because GREA does not explicitly decode or encode the subgraphs but directly

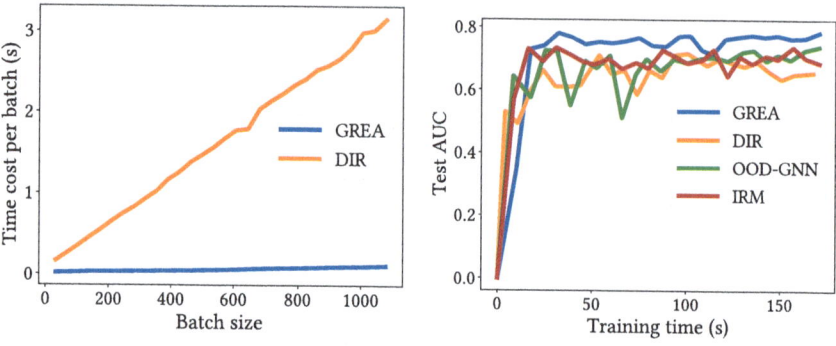

(a) GREA runs much faster than DIR when batch size (# graphs) increases.

(b) GREA spends comparable amount of training time to deliver the highest AUC.

**Fig. 4.5** Efficiency analysis on the ogbg-HIV dataset

creates their representations in latent space. Figure 4.5b shows that compared to three most competitive baselines, GREA delivers the highest AUC by learning augmented examples, while spending comparable amount of time.

### 4.4.6 RQ5: Sensitivity Analysis

We conduct three series of sensitivity analyses. First, Fig. 4.6 shows that on four polymer datasets, the performance of GREA in terms of $R^2$ is insensitive to the hyperparameters $\alpha$ and $\beta$ in Eq. (4.16). Second, Fig. 4.4 shows that the performance is insensitive to rationale size $\gamma$ in Eq. (4.14). Third, on two polymer datasets and one of the most popular molecule datasets, Table 4.4 compares the effects of different choices of AGG($\cdot$) function that aggregates the representations of rationale and environment subgraphs. Sum pooling is generally the best choice.

## 4.5 Conclusion

This chapter introduced GREA to improve graph rationale identification using data augmentations, including environment replacement, for accurate and interpretable graph property prediction. There was an efficient framework that performed rationale-environment separation and representation learning on real and augmented examples in one latent space. Experiments on molecule and polymer datasets demonstrated its effectiveness and efficiency. Case studies have demonstrated the coherence of rationale substructures discovered by GREA. However, further research on better regularization these rationale subgraphs remains promising.

## 4.5 Conclusion

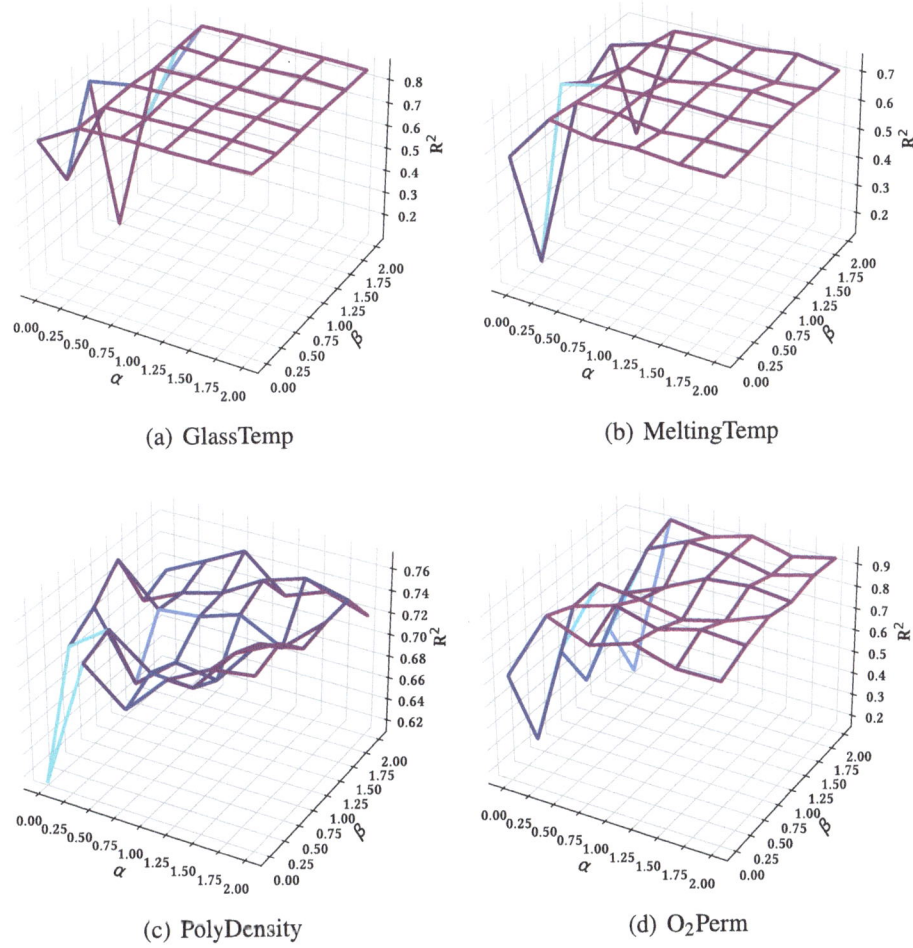

**Fig. 4.6** On four polymer datasets, the performance of GREA (in $R^2$) is *not* sensitive to hyperparameters $\alpha$ and $\beta$ in Eq. (4.16)

**Table 4.4** Effect of $\text{AGG}(\mathbf{h}_i^{(r)}, \mathbf{h}_j^{(e)})$ in Eq. (4.10). We use Sum Pooling by default because it generally performs the best

|  | MeltingTemp ($R^2$) | $O_2$Perm ($R^2$) | ogbg-HIV (AUC) |
|---|---|---|---|
| Sum pooling | **0.7362 ± 0.0115** | **0.9304 ± 0.0202** | **0.7932 ± 0.0092** |
| Mean pooling | 0.7328 ± 0.0068 | 0.9288 ± 0.0331 | 0.7810 ± 0.0117 |
| Max pooling | 0.7164 ± 0.0094 | 0.8984 ± 0.0494 | 0.7809 ± 0.0137 |
| Concatenation | 0.7145 ± 0.0127 | 0.9240 ± 0.0143 | 0.7771 ± 0.0096 |

# References

M. Arjovsky, L. Bottou, I. Gulrajani, and D. Lopez-Paz. Invariant risk minimization. *arXiv preprint* arXiv:1907.02893, 2019.

S. Chang, Y. Zhang, M. Yu, and T. Jaakkola. Invariant rationalization. In *International Conference on Machine Learning*, pages 1448–1458. PMLR, 2020.

L. Chen, G. Pilania, R. Batra, T. D. Huan, C. Kim, C. Kuenneth, and R. Ramprasad. Polymer informatics: Current status and critical next steps. *Materials Science and Engineering: R: Reports*, 144:100595, 2021.

S. Fan, X. Wang, C. Shi, P. Cui, and B. Wang. Generalizing graph neural networks on out-of-distribution graphs. *arXiv preprint* arXiv:2111.10657, 2021.

H. Gao and S. Ji. Graph u-nets. *IEEE Transactions on Pattern Analysis and Machine Intelligence*, 2021.

Z. Guo, C. Zhang, W. Yu, J. Herr, O. Wiest, M. Jiang, and N. V. Chawla. Few-shot graph learning for molecular property prediction. In *Proceedings of the Web Conference 2021*, pages 2559–2567, 2021.

W. L. Hamilton, R. Ying, and J. Leskovec. Inductive representation learning on large graphs. In *Proceedings of the 31st International Conference on Neural Information Processing Systems*, pages 1025–1035, 2017.

W. Hu, M. Fey, M. Zitnik, Y. Dong, H. Ren, B. Liu, M. Catasta, and J. Leskovec. Open graph benchmark: Datasets for machine learning on graphs. *Neural Information Processing Systems (NeurIPS)*, 2020.

C. Kim, A. Chandrasekaran, T. D. Huan, D. Das, and R. Ramprasad. Polymer genome: a data-powered polymer informatics platform for property predictions. *The Journal of Physical Chemistry C*, 122(31):17575–17585, 2018.

T. N. Kipf and M. Welling. Semi-supervised classification with graph convolutional networks. In *International Conference on Learning Representations*, 2017.

J. Lee, I. Lee, and J. Kang. Self-attention graph pooling. In *International Conference on Machine Learning*, pages 3734–3743. PMLR, 2019.

H. Li, X. Wang, Z. Zhang, and W. Zhu. Ood-gnn: Out-of-distribution generalized graph neural network. *arXiv preprint* arXiv:2112.03806, 2021.

R. Ma and T. Luo. Pi1m: a benchmark database for polymer informatics. *Journal of Chemical Information and Modeling*, 60(10):4684–4690, 2020.

R. Ma, Z. Liu, Q. Zhang, Z. Liu, and T. Luo. Evaluating polymer representations via quantifying structure–property relationships. *Journal of chemical information and modeling*, 59(7):3110–3119, 2019.

D. Mesquita, A. Souza, and S. Kaski. Rethinking pooling in graph neural networks. *Advances in Neural Information Processing Systems*, 33, 2020.

S. Otsuka, I. Kuwajima, J. Hosoya, Y. Xu, and M. Yamazaki. Polyinfo: Polymer database for polymeric materials design. In *2011 International Conference on Emerging Intelligent Data and Web Technologies*, pages 22–29. IEEE, 2011.

Y. Rong, W. Huang, T. Xu, and J. Huang. Dropedge: Towards deep graph convolutional networks on node classification. *arXiv preprint* arXiv:1907.10903, 2019.

E. Rosenfeld, P. K. Ravikumar, and A. Risteski. The risks of invariant risk minimization. In *International Conference on Learning Representations*, 2021. URL https://openreview.net/forum?id=BbNIbVPJ-42.

# References

D. F. Sanders, Z. P. Smith, R. Guo, L. M. Robeson, J. E. McGrath, D. R. Paul, and B. D. Freeman. Energy-efficient polymeric gas separation membranes for a sustainable future: A review. *Polymer*, 54(18):4729–4761, 2013.

A. Thornton, L. Robeson, B. Freeman, and D. Uhlmann. Polymer gas separation membrane database, 2012. URL https://research.csiro.au/virtualscreening/membrane-database-polymer-gas-separation-membranes/.

P. Veličković, G. Cucurull, A. Casanova, A. Romero, P. Liò, and Y. Bengio. Graph attention networks. In *International Conference on Learning Representations*, 2018.

Y. Wang, W. Wang, Y. Liang, Y. Cai, and B. Hooi. Graphcrop: Subgraph cropping for graph classification. *arXiv preprint* arXiv:2009.10564, 2020a.

Y. Wang, W. Wang, Y. Liang, Y. Cai, J. Liu, and B. Hooi. Nodeaug: Semi-supervised node classification with data augmentation. In *Proceedings of the 26th ACM SIGKDD International Conference on Knowledge Discovery & Data Mining*, pages 207–217, 2020b.

X. Wei, Z. Wang, Z. Tian, and T. Luo. Thermal transport in polymers: a review. *Journal of Heat Transfer*, 143(7):072101, 2021.

Y. Wu, X. Wang, A. Zhang, X. He, and T.-S. Chua. Discovering invariant rationales for graph neural networks. In *International Conference on Learning Representations*, 2022. URL https://openreview.net/forum?id=hGXij5rfiHw.

Z. Wu, B. Ramsundar, E. N. Feinberg, J. Gomes, C. Geniesse, A. S. Pappu, K. Leswing, and V. Pande. Moleculenet: a benchmark for molecular machine learning. *Chemical science*, 9(2):513–530, 2018.

Z. Wu, S. Pan, F. Chen, G. Long, C. Zhang, and S. Y. Philip. A comprehensive survey on graph neural networks. *IEEE transactions on neural networks and learning systems*, 32(1):4–24, 2020.

K. Xu, W. Hu, J. Leskovec, and S. Jegelka. How powerful are graph neural networks? In *International Conference on Learning Representations*, 2019. URL https://openreview.net/forum?id=ryGs6iA5Km.

J. Yang, L. Tao, J. He, J. McCutcheon, and Y. Li. Discovery of innovative polymers for next-generation gas-separation membranes using interpretable machine learning. *ChemRxiv*, 2021.

R. Ying, J. You, C. Morris, X. Ren, W. L. Hamilton, and J. Leskovec. Hierarchical graph representation learning with differentiable pooling. In *Proceedings of the 32nd International Conference on Neural Information Processing Systems*, pages 4805–4815, 2018.

R. Ying, D. Bourgeois, J. You, M. Zitnik, and J. Leskovec. Gnnexplainer: Generating explanations for graph neural networks. *Advances in neural information processing systems*, 32:9240, 2019.

Z. Zhang, P. Cui, and W. Zhu. Deep learning on graphs: A survey. *IEEE Transactions on Knowledge and Data Engineering*, 2020.

T. Zhao, Y. Liu, L. Neves, O. Woodford, M. Jiang, and N. Shah. Data augmentation for graph neural networks. In *Proceedings of the AAAI Conference on Artificial Intelligence*, volume 35, pages 11015–11023, 2021.

J. Zhou, J. Shen, and Q. Xuan. Data augmentation for graph classification. In *Proceedings of the 29th ACM International Conference on Information & Knowledge Management*, pages 2341–2344, 2020.

# Imbalanced Learning: Semi-Supervised Graph Imbalanced Regression

## 5.1 Introduction

Predicting graph properties has attracted great attention from drug discovery (Ramakrishnan et al. 2014; Wu et al. 2018) and material design (Ma and Luo 2020; Yuan et al. 2021), because molecules and polymers are naturally graphs. Properties like density, melting temperature, and oxygen permeability are often in continuous value spaces (Ramakrishnan et al. 2014; Wu et al. 2018; Yuan et al. 2021). Graph regression tasks are important and challenging. It is hard to observe label values in certain rare areas since the annotated data usually concentrate on small yet popular areas in the property spaces. Graph regression datasets are ubiquitously imbalanced. Previous attempts that address data imbalance mostly focused on categorical properties and classification tasks, however, *imbalanced regression tasks on graphs are under-explored.*

Besides data imbalance, the annotated graph regression data are often small in real world. For example, measuring the property of a molecule or polymer often needs expensive experiments or simulations. It has taken nearly 70 years to collect *only around 600* polymers with experimentally measured oxygen permeability in the Polymer Gas Separation Membrane Database (Thornton et al. 2012). On the other side, we have *hundreds of thousands* of unlabeled graphs.

Pseudo-labeling unlabeled graphs may enrich and balance training data, however, there are two challenges. First, if one directly trained a model on the imbalanced labeled data and used it to do pseudo-labeling, it is not reliable to generate accurate and balanced labels. Second, quite a number of unlabeled graphs might not follow the distribution of labeled data. Massive label noise is inevitable in pseudo-labeling. Therefore, selection is necessary to expand the set of data examples for training. Moreover, the selected pseudo-labels without noise cannot alleviate the label imbalance problem. Because the biased model tends to generate more pseudo-labels in the label ranges where most data concentrate. In this situation,

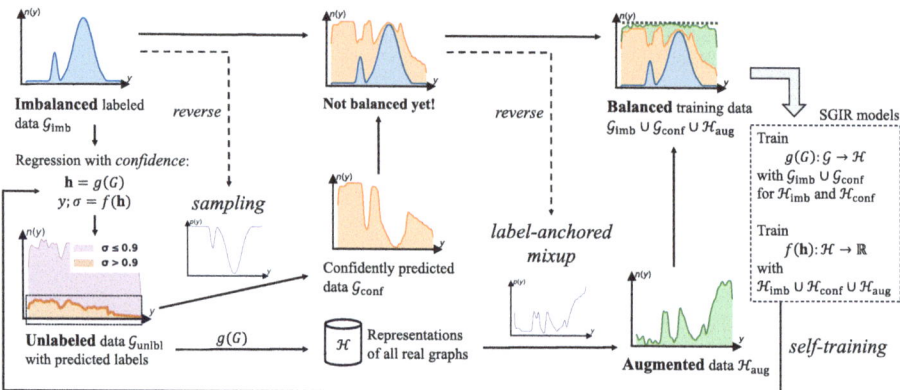

**Fig. 5.1** An overview of SGIR framework to train effective graph regression models with imbalanced labeled data. To balance the data properly, SGIR selects highly confident examples from predicted labels of unlabeled data and augments label areas that seriously lack data (even after added the confidently predicted data) by a novel label-anchored mixup algorithm

the selected pseudo-labels may aggravate the model bias and lead the model to have even worse performance on the label ranges where we lack enough data. Even though the pseudo-labeling had involved quality selection and the unlabeled set had been fully used to address label imbalance, the label distribution of annotated and pseudo-labeled examples might still be far from a perfect balance. That is, there might not be a sufficient number of pseudo-labeled examples to fill the gap in the under-represented label ranges.

Figure 5.1 illustrates ideas to overcome the above challenges. First, we want to progressively reduce the model bias by gradually improving training data from the labeled and unlabeled sets. The performance of pseudo-labeling models and the quality of the expanded training data can mutually enhance each other through iterations. Second, we relate the regression confidence to the prediction variance under perturbations. Higher confidence indicates a lower prediction variance under different perturbation environments. Therefore, we define and use *regression confidence* score to avoid pseudo-label noise and select quality examples in regression tasks. To fully exploit the quality pseudo-labels to compensate for the data imbalance in different label ranges, we use a reversed distribution of the imbalanced annotated data to reveal label ranges that need to be more or less selected for label balancing. Third, we attempt to achieve the perfect balance of training data by creating graph examples of any given label value in the remaining under-represented ranges.

In this chapter, we introduce SGIR, a Semi-supervised framework for Graph Imbalanced Regression. This framework has three novel designs to implement the ideas. First, SGIR is a self-training framework with multiple iterations for model learning and balanced training data generation. The second design is to sample more quality pseudo-labels for the less represented label ranges. We define a new measurement of regression confidence from recent studies on graph rationalization methods which provide perturbations for predictions at training and inference. After applying the confidence to filter out pseudo-label noise, we adopt *reverse sampling* to find optimal sampling rates at each label value that maximize the possibility of data balance. Intuitively, if a label value is less frequent in the annotated data, the sampling rate at this value is bigger and more pseudo-labeled examples are selected for model training. Third, we design a novel *label-anchored mixup* algorithm to augment graph examples by mixing up a virtual data point and a real graph example in latent space. Each virtual point is anchored at a certain label value that is still rare in the expanded labeled data. The mixed-up graph representations continue complementing the label ranges where we seriously lack data examples.

To empirically demonstrate the advantage of SGIR, we conduct experiments on seven graph property regression tasks from three different domains. Results show that SGIR significantly reduces the prediction error on all the tasks and in both under-/well-represented label ranges. For example, on the smallest dataset Mol-FreeSolv that has only 276 annotated graphs, SGIR reduces the mean absolute error from 1.114 to 0.777 (relatively 30% improvement) in the most under-represented label range and reduces the error from 0.642 to 0.563 (12% improvement) in the entire label space compared to state-of-the-art graph regression methods. To summarize:

- We focus a new problem of graph imbalance regression with a novel semi-supervised framework SGIR.
- SGIR is a self-training framework creating balanced and enriched training data from pseudo-labels and augmented examples with three collaborated components: regression confidence, reverse sampling, and label-anchored mixup.
- SGIR is theoretically motivated and empirically validated on seven graph regression tasks. It outperforms other semi-supervised learning and imbalanced regression methods in both well-represented and under-represented label ranges.

## 5.2 Problem Definition

To predict the property $y \in \mathbb{R}$ of a graph $G \in \mathcal{G}$, a graph regression model usually consists of an encoder $g : G \to \mathbf{h} \in \mathbb{R}^d$ and a decoder $f : \mathbf{h} \to \hat{y} \in \mathbb{R}$. The encoder $g(\cdot)$ is often a graph neural network (GNN) that outputs the $d$-dimensional representation vector $\mathbf{h}$ of graph $G$, and the decoder $f(\cdot)$ is often a multi-layer perceptron (MLP) that makes the label prediction $\hat{y}$ given $\mathbf{h}$.

Let $\mathcal{G}_{\text{imb}} = \{(G_i, y_i)\}_{i=1}^{n_{\text{imb}}}$ denote the labeled training data for graph regression models, where $n_{\text{imb}}$ is the number of training graphs in the imbalanced labeled dataset. It often concentrates on certain areas in the continuous label space. To reveal it, we first divide the label space into $C$ intervals and use them to fully cover the range of continuous label values. These intervals are $[b_0, b_1), [b_1, b_2), \ldots, [b_{C-1}, b_C)$. Then, we assign the labeled examples into $C$ intervals and count them in each interval to construct the frequency set $\{\mu_i\}_{i=1}^{C}$. We could find that $\frac{\max\{\mu_i\}}{\min\{\mu_i\}} \gg 1$ (*i.e.*, label imbalance) often exists, instead of $\mu_1 = \mu_2 = \cdots = \mu_C$ (*i.e.*, label balance) that is assumed by most existing models. The existing models may be biased to small areas in the label space that are dominated by the majority of labeled data and lack a good generalization to areas that are equally important but have much fewer examples.

Labeling continuous graph properties is difficult (Yuan et al. 2021), limiting the size of labeled data. Fortunately, a large number of unlabeled graphs are often available though ignored in most existing studies. In this chapter, we aim to use the unlabeled examples to alleviate the label imbalance issue in graph regression tasks. That is, let $\mathcal{G}_{\text{unlbl}} = \{G_j\}_{j=n_{\text{imb}}+1}^{n_{\text{imb}}+n_{\text{unlbl}}}$ denote the $n_{\text{unlbl}}$ available unlabeled graphs. We want to train $g(\cdot)$ and $f(\cdot)$ to deliver good performance through the whole continuous label space by utilizing both $\mathcal{G}_{\text{imb}}$ and $\mathcal{G}_{\text{unlbl}}$.

To progressively reduce label imbalance bias, we develop a novel framework named SGIR that iteratively creates reliable labeled examples in the areas of label space where annotations were not frequent. In Fig. 5.1, SGIR uses a graph regression model to create the labels and uses the gradually balanced data to train the regression model. To let data balancing and model construction mutually enhance each other, SGIR is a self-training framework that trains the encoder $g(\cdot)$ and decoder $f(\cdot)$ using two strategies through multiple iterations. The first strategy is to use pseudo-labels based on confident predictions and reverse sampling, leveraging unlabeled data (see Sect. 5.3.2). Because the unlabeled graph set still may not contain real examples of rare label values, the second strategy is to augment the graph representation examples for the rare areas using a novel label-anchored mixup algorithm (see Sect. 5.3.3).

## 5.3 Self-Training Framework: SGIR

### 5.3.1 A Self-Training Framework for Iteratively Balancing Scalar Label Data

A classic self-training framework is expected to be a virtuous circle exploiting the unlabeled data in label-balanced classification/regression tasks (Xie et al. 2020; McLachlan 1975). It first trains a classifier/regressor that iteratively assigns pseudo-labels to the set of unlabeled training examples $\mathcal{G}_{\text{unlbl}}$ with a margin greater than a certain threshold. The pseudo-labeled examples are then used to enrich the labeled training set. And the classifier continues training with the updated training set. For a virtuous circle of model training with imbalanced labeled set $\mathcal{G}_{\text{imb}}$, the most confident predictions on $\mathcal{G}_{\text{unlbl}}$ should be selected to compensate for the under-represented labels, as well as to enrich the dataset $\mathcal{G}_{\text{imb}}$. In each iteration, the model becomes less biased to the majority of labels. And the less biased model can make predictions of higher accuracy and confidence on the unlabeled data. Therefore, we hypothesize that model training and data balancing can mutually enhance each other.

## 5.3 Self-Training Framework: SGIR

SGIR is a self-training framework targeting to generalize the model performance everywhere in the continuous label space with particularly designed balanced training data from the labeled graph data $\mathcal{G}_{\text{imb}}$, confidently selected graph data $\mathcal{G}_{\text{conf}}$, and augmented representation data $\mathcal{H}_{\text{aug}}$. For the next round of model training, the gradually balanced training data reduce the label imbalance bias carried by the graph encoder $g(\cdot)$ and decoder $f(\cdot)$. Then the less biased graph encoder and decoder are applied to generate balanced training data of higher quality. Through these iterations, the model bias from the imbalanced or low-quality balanced data would be progressively reduced because of the gradually enhanced quality of balanced training data.

### 5.3.2 Balancing with Confidently Predicted Labels

At each iteration, SGIR enriches and balances training data with pseudo-labels of good quality. The unlabeled data examples in $\mathcal{G}_{\text{unlbl}}$ are firstly exploited by reliable and confident predictions. Then the reverse sampling from the imbalanced label distribution of original training data $\mathcal{G}_{\text{imb}}$ is used to select more pseudo-labels for under-represented label ranges.

**Graph Regression with Confidence**. A standard regression model outputs a scalar without a certain definition of confidence of its prediction. The confidence is often measured by how much the predicted probability is close to 1 in classifications. The lack of confidence measurements in graph regression tasks may introduce noise to the self-training framework that aims at label balancing. It would be more severe when the domain gap exists between labeled and unlabeled data (Berthelot et al. 2022). Recent studies Liu et al. (2022), Wu et al. (2022) have developed two concepts that help us define a good measurement: rationale subgraph and environment subgraph. A rationale subgraph is supposed to best support and explain the prediction at property inference. Its counterpart environment subgraph is the complementary subgraph in the example, which perturbs the prediction from the rationale subgraph if used. The idea is to measure the confidence of graph property prediction based on the reliability of the identified rationale subgraphs. Specifically, we use the variance of predicted label values from graphs that consist of a specific rationale subgraph and one of many possible environment subgraphs.

We use an existing supervised graph regression model that can identify rationale and environment subgraphs in any graph example to predict its property. We denote $G_i$ as the $i$-th graph in a batch of size $B$. The model separates $G_i$ into $G_i^{(r)}$ and $G_i^{(e)}$. For the $j$-th graph $G_j$ in the same batch, we have a combined example $G_{(i,j)} = G_i^{(r)} \cup G_j^{(e)}$ that has the rationale of $G_i$ and environment subgraph of $G_j$. So it is expected to have the same label of $G_i$. By enumerating $j \in \{1, 2, \ldots, B\}$, the encoder $g(\cdot)$ and decoder $f(\cdot)$ are trained to predict the label value of any $G_{(i,j)}$. We define the confidence of predicting the label of $G_i$ as:

$$\sigma_i = \frac{1}{\text{Var}\left(\{f(g(G_{(i,j)}))\}_{j=1,2,\ldots,B}\right)}. \tag{5.1}$$

It is the reciprocal of prediction variance. In implementation, we choose GREA (Liu et al. 2022) as the model. Considering efficiency, GREA creates $G_{(i,j)}$ in the latent space without decoding its graph structure. That is, it directly gets the representation of $G_{(i,j)}$ as the sum of the representation vectors $\mathbf{h}_i^{(r)}$ of $G_i^{(r)}$ and $\mathbf{h}_j^{(e)}$ of $G_j^{(e)}$. So we have

$$\sigma_i = \frac{1}{\text{Var}\left(\{f(\mathbf{h}_i^{(r)} + \mathbf{h}_j^{(e)})\}_{j=1,2,\ldots,B}\right)}. \tag{5.2}$$

Now we have predicted labels and confidence values for graph examples in the large unlabeled dataset $\mathcal{G}_{\text{unlbl}}$. Examples with low confidences will bring noise to the training data if we use them all. So we only consider a data example $G_i$ to be of good quality if its confidence $\sigma_i$ is not smaller than a threshold $\tau$. We name this confidence measurement based on graph rationalization as GRATION. GRATION is tailored for graph regression tasks by considering the environment subgraphs as perturbations. We will compare its effect on quality graph selection against other graph-irrelevant methods such as DROPOUT (Gal and Ghahramani 2016), CERTI (Tagasovska and Lopez-Paz 2019), DER (Deep Evidential Regression) (Amini et al. 2020), and SIMPLE (no confidence) in experiments.

After leveraging the unlabeled data, the label distribution of quality examples may still be biased to the majority of labels. So we further apply reverse sampling on these examples from $\mathcal{G}_{\text{unlbl}}$ to balance the distribution of training data.

**Reverse Sampling.** The reverse sampling in SGIR helps reduce the model bias to label imbalance. Specifically, we want to selectively add unlabeled examples predicted in the under-represented label ranges. Suppose we have the frequency set $\{\mu_i\}_{i=1}^C$ of $C$ intervals. We denote $p_i$ as the sampling rate at the $i$-th interval and follow Wei et al. (2021) to calculate it. Basically, to perform reverse sampling, we want $p_i < p_j$ if $\mu_i > \mu_j$. We define a new frequency set $\{\mu_i'\}_{i=1}^C$ in which $\mu_i'$ equals the $i$-th smallest in $\{\mu\}$ if $\mu_i$ is the $i$-th biggest in $\{\mu\}$. Then the sampling rate is

$$p_i = \frac{\mu_i'}{\max\{\mu_1, \mu_2, \ldots, \mu_C\}}. \tag{5.3}$$

To this end, we have the confidently labeled and reversed sampled data $\mathcal{G}_{\text{conf}}$. In each self-training iteration, we combine it with the original training set $\mathcal{G}_{\text{imb}}$.

### 5.3.3 Balancing with Augmentation via Label-Anchored Mixup

Although $\mathcal{G}_{\text{imb}} \cup \mathcal{G}_{\text{conf}}$ is more balanced than $\mathcal{G}_{\text{imb}}$, we observe that $\mathcal{G}_{\text{imb}} \cup \mathcal{G}_{\text{conf}}$ is usually far from a *perfect balance*, even if $\mathcal{G}_{\text{unlbl}}$ could be hundreds of times bigger than $\mathcal{G}_{\text{imb}}$. To create graph examples targeting the remaining under-represented label ranges, we design a novel label-anchored mixup algorithm for graph imbalanced regression. Compared to existing mixup algorithms (Wang et al. 2021; Verma et al. 2019) for classifications with-

## 5.3 Self-Training Framework: SGIR

out awareness of imbalance, the new algorithm can augment training data with additional examples for target ranges of continuous label value.

A mixup operation in the label-anchored mixup is to mix up two things in a latent space: (1) a virtual data point representing an interval of targeted label and (2) a real graph example. Specifically, we first calculate the representation of a target label interval by averaging the representation vectors of graphs in the interval from the labeled dataset $\mathcal{G}_{\text{imb}}$. Let $\mathbf{M} \in \{0, 1\}^{C \times n_{\text{imb}}}$ be an indicator matrix, where $M_{i,j} = 1$ means that the label of $G_j \in \mathcal{G}_{\text{imb}}$ belongs to the $i$-th interval. We denote $\mathbf{H} \in \mathbb{R}^{n_{\text{imb}} \times d}$ as the matrix of graph representations from the GNN encoder $g(\cdot)$ for $\mathcal{G}_{\text{imb}}$. The representation matrix $\mathbf{Z} \in \mathbb{R}^{C \times d}$ of all intervals is calculated

$$\mathbf{Z} = \text{norm}(\mathbf{M}) \cdot \mathbf{H}, \tag{5.4}$$

where $\text{norm}(\cdot)$ is the row-wise normalization and $a_i$ is the center value of the $i$-th interval. We have the representation-label pairs of all the label intervals $\{(\mathbf{z}_i, a_i)\}_{i=1}^{C}$, where $\mathbf{z}_i$ is the $i$-th row of $\mathbf{Z}$.

Now we can use each interval center $a_i$ as a label anchor to augment graph data examples in a latent space. We select $n_i \propto p_i$ real graphs from $\mathcal{G}_{\text{imb}} \cup \mathcal{G}_{\text{conf}}$ whose labels are closest to $a_i$, where $p_i$ is calculated by Eq. (5.3). The more real graphs are selected, the more graph representations are augmented. $n_i$ is likely to be big when the label anchor $a_i$ remains under-represented after $\mathcal{G}_{\text{conf}}$ is added to training set. Note that the labels were annotated if the graphs were in $\mathcal{G}_{\text{imb}}$ and predicted if they were in $\mathcal{G}_{\text{unlbl}}$. For $j \in \{1, 2, \ldots, n_i\}$, we mix up the interval $(\mathbf{z}_i, a_i)$ and a real graph $(\mathbf{h}_j, y_j)$, where $\mathbf{h}_j$ and $y_i$ are the representation vector and the annotated or predicted label of the $j$-th graph, respectively. Then the mixup operation is defined as

$$\begin{cases} \tilde{\mathbf{h}}_{(i,j)} = \lambda \cdot \mathbf{z}_i + (1 - \lambda) \cdot \mathbf{h}_j, \\ \tilde{y}_{(i,j)} = \lambda \cdot a_i + (1 - \lambda) \cdot y_j, \end{cases} \tag{5.5}$$

where $\tilde{\mathbf{h}}_{(i,j)}$ and $\tilde{y}_{(i,j)}$ are the representation vector and label of the augmented graph, respectively. $\lambda = \max(\lambda', 1 - \lambda')$, $\lambda' \sim \text{Beta}(1, \beta)$, and $\beta$ is a hyperparameter. $\lambda$ is often closer to 1 because we want $\tilde{y}_{(i,j)}$ to be closer to the label anchor $a_i$. Let $\mathcal{H}_{\text{aug}}$ denote the set of representation vectors of all the augmented graphs. Combined with $\mathcal{G}_{\text{imb}}$ and $\mathcal{G}_{\text{conf}}$, we end up with a label-balanced training set for the next round of self-training.

### 5.3.4 Optimization

In each iteration of self-training, we jointly optimize the parameters of graph encoder $g(\cdot)$ and label predictor $f(\cdot)$ with a gradually balanced training set $\mathcal{G}_{\text{imb}} \cup \mathcal{G}_{\text{conf}} \cup \mathcal{H}_{\text{aug}}$.

We use the mean absolute error (MAE) as the regression loss. Specifically, for each $(G, y) \in \mathcal{G}_{\text{imb}} \cup \mathcal{G}_{\text{conf}}$, the loss is $\ell_{\text{imb+conf}} = \text{MAE}(f(g(G)), y)$. Given $(\mathbf{h}, y) \in \mathcal{H}_{\text{aug}}$, the loss is $\ell_{\text{aug}} = \text{MAE}(f(\mathbf{h}), y)$. So the total loss for SGIR is

$$\mathcal{L} = \sum_{(G,y) \in \mathcal{G}_{imb} \cup \mathcal{G}_{conf}} \ell_{imb+conf}(G, y) + \sum_{(\mathbf{h},y) \in \mathcal{H}_{aug}} \ell_{aug}(\mathbf{h}, y).$$

The framework is flexible with any graph encoder-decoder models. To be consistent and given the promising results in graph regression tasks, we use the design of graph encoder and decoder in GREA (Liu et al. 2022) which is also used for measuring prediction confidence in Eq. (5.2).

## 5.4 Theoretical Motivations

There is a lack of theoretical principle for imbalanced regression. The theoretical motivation extends the generalization error bound from classification (Cao et al. 2019) to regression. The original bound enforces bigger margins for minority classes, which potentially hurt the model performance for well-represented classes (Tian et al. 2020; Zhang et al. 2023). The result provides a more safe way to reduce the error bound by utilizing unlabeled graphs with self-training in graph regression tasks.

As we divide the label distribution into $C$ intervals, every graph example can be assigned into an interval (as the ground-truth interval) according to the distance between the interval center and the ground-truth label value. Besides, we use $S_{[b_i,b_{i+1})}(G)$ to denote the reciprocal of the distance between the predicted label of the graph $G$ and the $i$-th interval $[b_i, b_{i+1})$, where $i \in \{1, 2, \ldots, C\}$. In this way, we could define $f(\cdot)$ as a regression function that outputs a continuous predicted label. Then $S_{[b_i,b_{i+1})}(G)$ consists of $f(\cdot)$ and outputs the logits to classify the graph to the $i$-the interval.

We consider all training examples to follow the same distribution. We assume that conditional on label intervals, the distributions of graph sampling are the same at training and testing stages. So, the standard 0–1 test error on the balanced test distribution is

$$\mathcal{E}_{bal}[f] = \Pr_{(G,[b_i,b_{i+1})) \sim \mathcal{P}_{bal}} \left[ S_{[b_i,b_{i+1})}(G) < \max_{j \neq i} S_{[b_j,b_{j+1})}(G) \right], \quad (5.6)$$

where $\mathcal{P}_{bal}$ denotes the balanced test distribution. It first samples a label interval uniformly and then samples graphs conditionally on the interval. The error for the $i$-th interval $[b_i, b_{i+1})$ is defined as

$$\mathcal{E}_{[b_i,b_{i+1})}[f] = \Pr_{G \sim \mathcal{P}_{[b_i,b_{i+1})}} \left[ S_{[b_i,b_{i+1})}(G) < \max_{j \neq i} S_{[b_j,b_{j+1})}(G) \right], \quad (5.7)$$

where $\mathcal{P}_{[b_i,b_{i+1})}$ denotes the distribution for the interval $[b_i, b_{i+1})$. We define $\gamma(G, [b_i, b_{i+1})) = S_{[b_i,b_{i+1})}(G) - \max_{j \neq i} S_{[b_j,b_{j+1})}(G)$ as the margin of an example $G$ assigned to the interval $[b_i, b_{i+1})$. To define the training margin $\gamma_{[b_i,b_{i+1})}$ for the interval $[b_i, b_{i+1})$, we calculate the minimal margin across all examples assigned to that interval:

## 5.4 Theoretical Motivations

$$\gamma_{[b_i,b_{i+1})} = \min_{G_j \in [b_i,b_{i+1})} \gamma\left(G_j, [b_i, b_{i+1})\right). \tag{5.8}$$

We assume that the MAE regression loss is small enough to correctly assign all training examples to the corresponding intervals. Given the hypothesis class $\mathcal{F}$, $C(\mathcal{F})$ is assumed to be a proper complexity measure of $\mathcal{F}$. We assume there are $n_{[b_i,b_{i+1})}$ examples i.i.d sampled from the conditional distribution $\mathcal{P}_{[b_i,b_{i+1})}$ for the interval $[b_i, b_{i+1})$. So, we apply the standard margin-based generalization bound to obtain the following theorem (Kakade et al. 2008; Cao et al. 2019; Zhao et al. 2021):

**Theorem 5.1** *With probability $(1 - \delta)$ over the randomness of the training data, the error $\mathcal{E}_{[b_i,b_{i+1})}$ for interval $[b_i, b_{i+1})$ is bounded by*

$$\mathcal{E}_{[b_i,b_{i+1})}[f] \lesssim \frac{1}{\gamma_{[b_i,b_{i+1})}} \sqrt{\frac{C(\mathcal{F})}{n_{[b_i,b_{i+1})}}} \\ + \sqrt{\frac{\log\log_2(1/\gamma_{[b_i,b_{i+1})}) + \log(1/\delta)}{n_{[b_i,b_{i+1})}}}, \tag{5.9}$$

*where $\lesssim$ hides constant terms. Taking union bound over all intervals, we have $\mathcal{E}_{\text{bal}}[f] \lesssim \frac{1}{C} \sum_{i=1}^{C} \mathcal{E}_{[b_i,b_{i+1})}[f]$.*

We rely on two theorems to derive Theorem 5.1.

**Existing Theorems**
Given a classifier $f$ from the function class $\mathcal{F}$, an input example $x$ from the feature space $\mathcal{X}$ and its label $y$.

**Theorem 5.2** (from Bartlett and Mendelson (2002), Kakade et al. (2008)) *Assume the expected loss on examples is $\mathcal{E}[f]$ and the corresponding empirical loss $\hat{\mathcal{E}}[f]$. Assume the loss is Lipschitz with Lipschitz constant $L_e$. And it is bounded by $c_0$. For any $\delta > 0$ and with probability at least $1 - \delta$ simultaneously for all $f \in \mathcal{F}$ we have that*

$$\mathcal{E}[f] \leq \hat{\mathcal{E}}[f] + 2L_e \mathcal{R}_n(\mathcal{F}) + c_0 \sqrt{\frac{\log(1/\delta)}{2n}}, \tag{5.10}$$

*where $n$ is the number of example and $\hat{\mathcal{R}}_n(\mathcal{F})$ is the Rademacher complexity measurement of the hypothesis class $\mathcal{F}$.*

**Theorem 5.3** (from Kakade et al. (2008)) *Applying Theorem 5.2 and considering the fraction of data having $\gamma$-margin mistakes, or $K_\gamma[f] := \frac{|i:y_i f(x_i) < \gamma|}{n}$. Assume $\forall f \in \mathcal{F}$ we have $\sup_{x \in \mathcal{X}} |f(x)| \leq c_1$. Then, with probability at least $1 - \delta$ over the example, for all margins $\gamma > 0$ and all $f \in \mathcal{F}$ we have,*

$$\mathcal{E}[f] \leq K_\gamma[f] + 4\frac{\mathcal{R}_n(\mathcal{F})}{\gamma} + \sqrt{\frac{2\log\left(\log_2(4c_1/\gamma)\right) + \log(1/\delta)}{2n}}, \quad (5.11)$$

$$\leq K_\gamma[f] + 4\frac{\mathcal{R}_n(\mathcal{F})}{\gamma} + \sqrt{\frac{\log\left(\log_2 \frac{4c_1}{\gamma}\right)}{n}} + \sqrt{\frac{\log(1/\delta)}{2n}}. \quad (5.12)$$

Proof of Theorem 5.1

In this chapter, we use the regression function $f$ to predict the label value. We calculate the reciprocal of the distance between the predicted label and interval centers as unnormalized probabilities of the graph $S_{[b_i,b_{i+1})}(G)$ being assigned to the interval $[b_i, b_{i+1}), i \in \{1, 2, \ldots, C\}$. Given a hard margin $\gamma$, we use $\mathcal{E}_{\gamma,[b_i,b_{i+1})}[f]$ to denote the hard margin loss for examples in the interval $[b_i, b_{i+1})$:

$$\mathcal{E}_{\gamma,[b_i,b_{i+1})}[f] = \Pr_{G \sim \mathcal{P}_{[b_i,b_{i+1})}} \left[ S_{[b_i,b_{i+1})}(G) < \max_{j \neq i} S_{[b_j,b_{j+1})}(G) + \gamma \right]. \quad (5.13)$$

We assume its empirical variant is $\hat{\mathcal{E}}_{\gamma,[b_i,b_{i+1})}[f]$. The empirical Rademacher complexity $\hat{\mathcal{R}}_{(b_i,b_{i+1}]}(\mathcal{F})$ is used as the complexity measurement $C(\mathcal{F})$ for the hypothesis class $\mathcal{F}$. With a vector $\sigma$ of i.i.d. uniform $\{-1, +1\}$ bits, we have

$$\hat{\mathcal{R}}_{(b_i,b_{i+1}]}(\mathcal{F}) = \quad (5.14)$$

$$\frac{1}{n_{(b_i,b_{i+1}]}} \mathbb{E}_\sigma \left[ \sup_{f \in \mathcal{F}} \sum_{G_i \in [b_i,b_{i+1})} \sigma_i \left[ S_{[b_i,b_{i+1})}(G_i) - \max_{j \neq i} S_{[b_j,b_{j+1})}(G_i) \right] \right] \quad (5.15)$$

As any $G_i$ in the interval $(b_i, b_{i+1}]$ is an i.i.d. sample from the distribution $\mathcal{P}_{[b_i,b_{i+1})}$, we directly apply the standard margin-based generalization bound Theorem 5.3 (Kakade et al. 2008): with probability $1 - \delta$, for all choices of $\gamma_{[b_i,b_{i+1})} > 0$ and $f \in \mathcal{F}$,

$$\mathcal{E}_{[b_i,b_{i+1})} \leq \hat{\mathcal{E}}_{\gamma,[b_i,b_{i+1})}[f] + 4\frac{\hat{\mathcal{R}}_{(b_i,b_{i+1}]}(\mathcal{F})}{\gamma_{[b_i,b_{i+1})}} \quad (5.16)$$

$$+ \sqrt{\frac{2\log\left(\log_2(\frac{4c_1}{\gamma_{[b_i,b_{i+1})}})\right) + \log(1/\delta)}{2n_{[b_i,b_{i+1})}}},$$

$$\leq \hat{\mathcal{E}}_{\gamma,[b_i,b_{i+1})}[f] + \frac{1}{\gamma_{[b_i,b_{i+1})}} \sqrt{\frac{C(\mathcal{F})}{n_{[b_i,b_{i+1})}}} \quad (5.17)$$

$$+ \sqrt{\frac{2\log\left(\log_2(\frac{4c_1}{\gamma_{[b_i,b_{i+1})}})\right) \log(1/\delta)}{2n_{[b_i,b_{i+1})}}},$$

## 5.4 Theoretical Motivations

$$\lesssim \frac{1}{\gamma_{[b_i,b_{i+1}]}} \sqrt{\frac{C(\mathcal{F})}{n_{[b_i,b_{i+1}]}}} + \sqrt{\frac{\log\log_2(1/\gamma_{[b_i,b_{i+1}]}) + \log(1/\delta)}{2n_{[b_i,b_{i+1}]}}}. \quad (5.18)$$

We derive Eq. (5.17) from Eq. (5.16) because the Rademacher complexity $\hat{\mathcal{R}}_{(b_i,b_{i+1}]}(\mathcal{F})$ typically scales as $\sqrt{\frac{C(\mathcal{F})}{n_{(b_i,b_{i+1}]}}}$ for some complexity measurement $C(\mathcal{F})$ (Cao et al. 2019). We derive Eq. (5.18) from Eq. (5.17) by ignoring constant factors (Cao et al. 2019). Since the overall performance $\mathcal{E}_{\text{bal}}[f]$ is calculated over all intervals, we get it as $\mathcal{E}_{\text{bal}}[f] = \frac{1}{C}\sum_{i=1}^{C}\mathcal{E}_{[b_i,b_{i+1}]}$.

The bound decreases as the increase of the examples in corresponding label ranges. The SGIR is motivated to reduce and balance the bound for different intervals by manipulating $n_{[b_i,b_{i+1}]}$ with pseudo-labels and augmented examples. In the following, we discuss that the augmented examples do not break the assumption for the theorem and future directions of imbalanced regression theories without intervals.

### Discussions

Existing work on the theoretical analysis of mixup (Liu et al. 2023; Carratino et al. 2020; Zhang et al. 2021) mainly focused on image classification and the augmented graph examples are not i.i.d sampled from the conditional distribution $\mathcal{P}_{[b_i,b_{i+1}]}$ for a specific interval $[b_i, b_{i+1}]$. While a recent work developed the C-mixup (Yao et al. 2022) to sample closer pairs of examples with higher probability for regression tasks, it did not fit the theoretical motivation to address the label imbalance issue: with C-mixup, the pairs in the over-represented label ranges have a higher probability to be sampled than the under-represented ones. Compared to these theories and methods for the mixup algorithm, the label-anchored mixup allows direct application to imbalanced regression tasks without compromising the assumption in the theoretical motivation. This is because we use the augmented virtual examples $\mathcal{H}_{\text{aug}}$ based on the label anchor within intervals $[b_i, b_{i+1}]$. Augmented examples are independently created with Eq. (5.5). Since the interval centers could be mixed with any other real graphs from $\mathcal{G}_{\text{imb}} \cup \mathcal{G}_{\text{conf}}$, any value in the interval space could be sampled. Besides, it is reasonable to use the distribution of the entire label space (from $\mathcal{G}_{\text{imb}} \cup \mathcal{G}_{\text{conf}}$) to approximate the distribution within the interval and assume that the conditional distribution $\mathcal{P}_{[b_i,b_{i+1}]}$ does not change.

We present the theoretical principle for imbalanced regression with intervals to connect with existing theoretical principles for classification. Future theoretical work on imbalanced regression can leverage the advantages of using mixture regressor models (Sugiyama and Storkey 2006), which have been used to address covariate shift problems in regression tasks. Additionally, exploring the promising connection between domain adaptation theories and sample selection bias (Sun et al. 2016; Cortes et al. 2008) holds potential in this field.

## 5.5 Experiments

We present experiments to demonstrate the effectiveness of SGIR and answer the research question: how it performs on graph regression tasks and at different label ranges (RQ1). We also make a few ablation studies to investigate the effect of model design: where the effectiveness comes from (RQ2).

### 5.5.1 Experimental Settings

**Datasets**. We give a comprehensive introduction to the datasets used for regression tasks and splitting idea from Gong et al. (2022), Yang et al. (2021). *Mol-Lipo* predicts the property of lipophilicity consisting of 4200 molecules. The lipophilicity is important for solubility and membrane permeability in drug molecules. This dataset originates from ChEMBL (Mendez et al. 2019). The property is from experimental results for the octanol/water distribution coefficient (log $D$ at pH 7.4). *Mol-ESOL* predicts the water solubility (log solubility in mols per litre) from chemical structures consisting of 1128 small organic molecules. *Mol-FreeSolv* predicts the hydration free energy of molecules in water consisting of 642 molecules. The property is experimentally measured or calculated. *MeltingTemp* predicts the property of melting temperature (°C). It is collected from PolyInfo, a web-based polymer database (Otsuka et al. 2011). *PolyDensity* predicts the property of polymer density (g/cm$^3$). It is collected from PolyInfo, a web-based polymer database (Otsuka et al. 2011). $O_2$*Perm* predicts the property of oxygen permeability (Barrer). It is created from the Membrane Society of Australasia portal consisting of experimentally measured gas permeability data (Thornton et al. 2012). *Unlabeled Data for Molecules and Polymers* The total number of unlabeled graphs for molecule and polymers is 146,129, consisting of 133,015 molecules from QM9 (Ramakrishnan et al. 2014) and 13,114 monomers (the repeated units of polymers) from Liu et al. (2022). QM9 is a molecule dataset for stable small organic molecules consisting of atoms C, H, O, N, and F. We use it as a source of unlabeled data. We integrate four polymer regression datasets including MeltingTemp, PolyDensity, $O_2$Perm and another one from Liu et al. (2022) for the glass transition temperature as the other source of unlabeled data. The unlabeled graphs may be slightly less than 146,129 for a polymer task on MeltingTemp, PolyDensity or $O_2$Perm. It is because we remove the overlapping graphs for the current polymer task with the polymer unlabeled data.

**Data splitting for Molecules and Polymers**. We split the datasets based on the approach in previous works Gong et al. (2022), Yang et al. (2021) motivated for two reasons. First, we want the training sets to well characterize the imbalanced label distribution as presented in the original datasets. Second, we want relatively balanced valid and test sets to fairly evaluate the model performance in different ranges of label values.

## 5.5 Experiments

**Table 5.1** Statistics of six tasks for graph property regression

| Dataset | # Graphs (train/valid/test) | # Nodes (avg./max) | # Edges (avg./max) |
|---|---|---|---|
| Mol-Lipo | 2,048/1,076/1,076 | 27.0/115 | 59.0/236 |
| Mol-ESOL | 446/341/341 | 13.3/55 | 27.4/125 |
| Mol-FreeSolv | 276/183/183 | 8.7/24 | 16.8/50 |
| MeltingTemp | 2,419/616/616 | 26.9/102 | 55.4/212 |
| PolyDensity | 844/425/425 | 27.3/93 | 57.6/210 |
| $O_2$Perm | 339/128/128 | 37.3/103 | 82.1/234 |
| Superpixel-Age | 3619/628/628 | 67.9/75.0 | 265.6/300 |

**Superpixel-Age.** The details of the age regression dataset are presented in Table 5.1 (Superpixel-Age) and Fig. 5.3. The graph dataset Superpixel-Age is constructed from image superpixels using the algorithms from Knyazev et al. (2019) on the image dataset *AgeDB-DIR* from Moschoglou et al. (2017), Yang et al. (2021). Each face image in *AgeDB-DIR* has an age label from 0 to 101. We fisrt compute the SLIC superpixels for each image without losing the label-specific information (Achanta et al. 2012; Knyazev et al. 2019). Then we use the superpixels as nodes and calculate the spatial distance between superpixels to build edges for each image (Knyazev et al. 2019). Binary edges are constructed between superpixel nodes by applying a threshold on the top-5% of the smallest spatial distances. After building a graph for each image, we follow the data splitting in Yang et al. (2021) to study the imbalanced regression problem. We randomly remove 70% labels in the training/validation/test data and use them as unlabeled graphs. Finally, the graph dataset Superpixel-Age consists of 3,619 graphs for training, 628 graphs for validation, 628 graphs for testing, and 11,613 unlabeled graphs for semi-supervised learning.

**Evaluation metrics.** We report model performance on three different sub-ranges following the work in Gong et al. (2022), Ren et al. (2022), Yang et al. (2021), besides the *entire range* of label space. The three sub-ranges are the *many-shot region*, *medium-shot region*, and *few-shot region*. The sub-ranges are defined by the number of training graphs in each label value interval. Details for each dataset are presented in Figs. 5.2 and 5.3. To evaluate the regression performance, we use mean absolute error (MAE) and geometric mean (GM) (Yang et al. 2021). Lower values (↓) of MAE or GM indicate better performance.

**Baselines and Implementations.** Besides the GNN model, we broadly consider baselines from the fields of imbalanced regression and semi-supervised graph learning. Specifically, imbalanced regression baselines include LDS (Yang et al. 2021), BMSE (Ren et al. 2022), and RANKSIM (Gong et al. 2022). The semi-supervised graph learning baseline is INFOGRAPH (Sun et al. 2020) and the graph learning baseline is GREA (Liu et al. 2022).

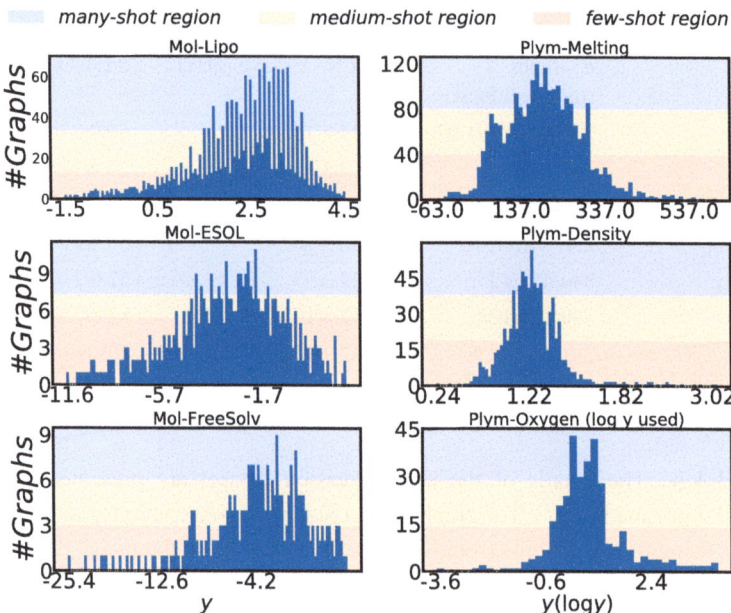

**Fig. 5.2** Imbalanced training distributions $\mathcal{G}_{\text{imb}}$ of labeled molecule and polymers

**Fig. 5.3** Imbalanced training distributions $\mathcal{G}_{\text{imb}}$ in the Superpixel-Age dataset

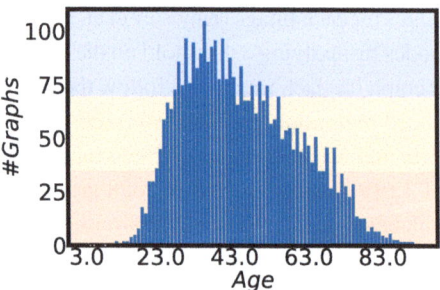

**SGIR Implementation.** SGIR utilizes GIN (Xu et al. 2019) as the GNN encoder and the decoder is a three-layer MLP to output property values. The threshold $\tau$ for selecting confident predictions is determined by the value at a certain percentile of the confidence score distribution. We use the Graph Isomorphism Network (GIN) (Xu et al. 2019) as the GNN encoder for $f_\theta$ to get the graph representation and three layers of Multilayer perceptron (MLP) as the decoder to predict graph properties. The threshold $\tau$ for selecting confident predictions is determined by the value at a certain percentile of the confidence score distribution. To implement it, we set it up as a hyperparameter $\tau_{\text{pct}}$ determining the percentile value of the prediction variance (i.e., the reciprocal of confidence) of the labeled training data. All methods are implemented on a RTX 2080Ti card (11 GB RAM). We

## 5.5 Experiments

reports the results on the test sets using the mean (standard deviation) over 10 runs. Note that the underlying design of the graph learning model used in SGIR is GREA with a learning objective as follows. Given $(G, y) \in \mathcal{G}_{\text{imb}} \cup \mathcal{G}_{\text{conf}}$, GREA (Liu et al. 2022) will output a vector $\mathbf{m} \in \mathbb{R}^K$ that indicates the probability of $K$ nodes in a graph being in the rationale subgraph. So, we could get $\mathbf{h}^{(r)} = \mathbf{1}_K^\top \cdot (\mathbf{m} \times \mathbf{H})$ and $\mathbf{h}^{(e)} = \mathbf{1}_K^\top \cdot ((\mathbf{1}_K - \mathbf{m}) \times \mathbf{H})$, where $\mathbf{H} \in \mathbb{R}^{K \times d}$ is the node representation matrix. By this, the optimization objectives of a graph consist of

$$\begin{cases} \ell_{\text{imb+conf}} = \text{MAE}(f(\mathbf{h}^{(r)}), y) + \mathbb{E}_{G'}\big[\text{MAE}(f(\mathbf{h}+\mathbf{h}'), y)\big] \\ \qquad\qquad + \text{Var}_{G'}\big(\{\text{MAE}(f(\mathbf{h}+\mathbf{h}'), y)\}\big), \\ \ell_{\text{regu}} = \frac{1}{K}\sum_{k=1}^{K} |m_k| - \gamma \end{cases}$$

$\ell_{\text{regu}}$ regularizes the vector $\mathbf{m}$ and $\gamma \in [0, 1]$ is a hyperparameter to control the expected size of $G^{(r)}$. $G'$ is the possible graph in the same batch that provides environment subgraphs and $\mathbf{h}'$ is the representation vector of the environment subgraph. When combining the rationale-environment pairs to create new graph examples, the original GREA creates the same number of examples for the under-represented rationale and the well/over-represented rationale. We observe that it may make the training examples more imbalanced. Therefore, we use the reweighting technique to penalize more for the expectation term ($\mathbb{E}_{G'}\big[\text{MAE}(f(\mathbf{h}+\mathbf{h}'), y)\big]$) and variance term ($\text{Var}_{G'}\big(\{\text{MAE}(f(\mathbf{h}+\mathbf{h}'), y)\}\big)$) in $\ell_{\text{imb+conf}}$ when the label is from the under-represented ranges. The weight of the expectation and variance terms for a graph with label $y$ is

$$w = \frac{\exp(\sum_{b=1}^{B} |y - y_b|/t)}{\exp(\sum_{j=1}^{B} \sum_{b=1}^{B} |y - y_b|/t)},$$

where $B$ is the batch size and $t$ is the temperature hyper-parameter.

For all the methods, we reports the results on the test sets using the mean (standard deviation) over 10 runs with parameters that are randomly initialized.

### 5.5.2 RQ1: Effectiveness on Property Prediction

Table 5.2 presents results of SGIR and baseline methods on six graph regression tasks. We have three observations.

**Overall effectiveness in the entire label range**: SGIR performs consistently better than competitive baselines on all tasks. Columns "All" in Table 5.2 show that SGIR reduces MAE over the best baselines (whose MAEs are underlined in the table) relatively by 9.1%, 8.1%, and 12.3% on the three molecule datasets, respectively. Specifically, on Mol-FreeSolv, the MAE was reduced from 0.642 to 0.563 with no change on the standard deviation. This is because SGIR enriches and *balances* the training data with confidently predicted pseudo-

**Table 5.2** Results of MEAN(STD) on six molecule/polymer datasets

| | | MAE ↓ | | | | GM ↓ | | | |
|---|---|---|---|---|---|---|---|---|---|
| | | All | Many-shot | Med.-shot | Few-shot | All | Many-shot | Med.-shot | Few-shot |
| Mol-Lipo | GNN | 0.485(0.010) | 0.421(0.030) | 0.462(0.013) | 0.566(0.032) | 0.297(0.012) | 0.252(0.022) | 0.294(0.016) | 0.348(0.030) |
| | RANKSIM | 0.475(0.018) | 0.388(0.017) | 0.438(0.007) | 0.587(0.043) | 0.297(0.015) | 0.249(0.017) | 0.274(0.006) | 0.380(0.044) |
| | BMSE | 0.494(0.007) | 0.409(0.019) | 0.450(0.007) | 0.614(0.033) | 0.304(0.008) | 0.260(0.014) | 0.279(0.015) | 0.382(0.038) |
| | LDS | 0.468(0.009) | 0.394(0.012) | 0.449(0.012) | 0.551(0.026) | 0.294(0.010) | 0.251(0.009) | 0.281(0.010) | 0.356(0.033) |
| | INFOGRAPH | 0.499(0.008) | 0.421(0.024) | 0.471(0.013) | 0.596(0.026) | 0.314(0.011) | 0.269(0.018) | 0.300(0.006) | 0.376(0.029) |
| | GREA | 0.487(0.002) | 0.391(0.015) | 0.434(0.008) | 0.626(0.018) | 0.294(0.010) | 0.251(0.009) | 0.281(0.010) | 0.356(0.033) |
| | SGIR | **0.432(0.012)** | **0.357(0.019)** | **0.413(0.017)** | **0.515(0.020)** | **0.264(0.013)** | **0.224(0.016)** | **0.256(0.017)** | **0.314(0.015)** |
| Mol-ESOL | GNN | 0.508(0.015) | 0.398(0.018) | 0.448(0.012) | 0.696(0.025) | 0.299(0.017) | 0.231(0.017) | 0.279(0.014) | 0.425(0.035) |
| | RANKSIM | 0.501(0.014) | 0.389(0.021) | 0.443(0.019) | 0.689(0.025) | 0.293(0.021) | 0.227(0.028) | 0.258(0.020) | 0.449(0.030) |
| | BMSE | 0.533(0.023) | 0.400(0.027) | 0.449(0.015) | 0.777(0.069) | 0.308(0.018) | 0.245(0.036) | 0.266(0.009) | 0.473(0.035) |
| | LDS | 0.517(0.016) | 0.423(0.012) | 0.474(0.029) | 0.668(0.010) | 0.304(0.010) | 0.261(0.007) | 0.283(0.025) | 0.393(0.009) |
| | INFOGRAPH | 0.561(0.025) | 0.475(0.034) | 0.466(0.036) | 0.776(0.036) | 0.336(0.014) | 0.306(0.022) | 0.276(0.013) | 0.484(0.029) |
| | GREA | 0.497(0.031) | 0.396(0.040) | 0.456(0.033) | 0.652(0.045) | 0.289(0.032) | 0.226(0.038) | 0.270(0.025) | 0.404(0.051) |
| | SGIR | **0.457(0.015)** | **0.370(0.022)** | **0.411(0.011)** | **0.604(0.024)** | **0.263(0.016)** | **0.226(0.021)** | **0.240(0.015)** | **0.347(0.030)** |
| Mol-FreeSolv | GNN | 0.726(0.039) | 0.617(0.061) | 0.695(0.055) | 1.154(0.082) | 0.363(0.025) | 0.317(0.027) | 0.360(0.029) | 0.556(0.073) |
| | RANKSIM | 0.779(0.109) | 0.764(0.225) | 0.674(0.072) | 1.220(0.146) | 0.367(0.026) | 0.396(0.052) | 0.315(0.030) | 0.537(0.082) |
| | BMSE | 0.856(0.071) | 0.809(0.117) | 0.820(0.064) | 1.122(0.076) | 0.456(0.042) | 0.426(0.029) | 0.457(0.054) | 0.552(0.062) |
| | LDS | 0.809(0.071) | 0.796(0.071) | 0.737(0.088) | 1.114(0.141) | 0.443(0.045) | 0.489(0.036) | 0.387(0.052) | 0.580(0.146) |
| | INFOGRAPH | 0.933(0.042) | 0.830(0.081) | 0.913(0.030) | 1.308(0.171) | 0.542(0.048) | 0.505(0.107) | 0.528(0.038) | 0.789(0.183) |
| | GREA | 0.642(0.026) | 0.541(0.064) | 0.570(0.008) | 1.202(0.023) | 0.321(0.038) | 0.294(0.064) | 0.301(0.024) | 0.537(0.049) |
| | SGIR | **0.563(0.026)** | **0.535(0.038)** | **0.528(0.046)** | **0.777(0.061)** | **0.264(0.029)** | **0.286(0.013)** | **0.244(0.046)** | **0.304(0.078)** |
| MeltingTemp | GNN | 41.8(1.2) | 35.5(1.2) | 33.0(0.7) | 54.7(2.2) | 23.2(1.0) | 21.3(1.1) | 16.2(1.0) | 33.4(2.5) |
| | RANKSIM | 41.1(0.9) | 34.1(0.5) | 33.6(1.1) | 53.5(1.2) | 22.6(1.1) | 20.5(0.5) | 16.8(1.0) | 31.4(2.8) |
| | BMSE | 42.1(0.7) | 35.8(1.4) | 34.1(1.3) | 54.4(1.5) | 23.7(1.2) | 21.5(1.0) | 18.1(0.5) | 32.4(3.0) |
| | LDS | 41.6(0.3) | 35.3(0.9) | 34.5(1.1) | 53.2(0.8) | 23.2(0.2) | 20.5(1.2) | 18.3(0.5) | 31.4(1.1) |
| | INFOGRAPH | 43.6(2.8) | 35.3(2.3) | 35.0(2.3) | 58.3(4.1) | 24.6(1.9) | 21.3(1.5) | 18.4(1.5) | 35.4(4.1) |
| | GREA | 41.2(0.8) | 33.3(0.5) | 32.7(0.7) | 55.3(3.0) | 23.4(0.6) | 20.0(0.6) | 17.3(0.7) | 34.3(2.9) |
| | SGIR | **38.9(0.7)** | **31.7(0.3)** | **31.5(1.1)** | **51.4(1.6)** | **21.1(1.2)** | **18.5(0.5)** | **15.9(1.4)** | **30.2(1.9)** |

(continued)

## 5.5 Experiments

Table 5.2 (continued)

| | | MAE ↓ | | | | GM ↓ | | | |
|---|---|---|---|---|---|---|---|---|---|
| | | All | Many-shot | Med.-shot | Few-shot | All | Many-shot | Med.-shot | Few-shot |
| PolyDensity (scaled:×1.000) | GNN | 61.2(5.4) | 63.4(18.9) | 46.6(1.6) | 72.0(2.8) | 29.3(0.6) | 29.6(3.3) | 23.5(0.9) | 35.5(2.0) |
| | RANKSIM | 57.5(1.8) | 55.1(2.2) | 46.3(1.8) | 69.4(3.3) | 29.3(1.6) | 29.9(2.8) | 23.1(2.1) | 35.4(2.5) |
| | BMSE | 61.8(2.0) | 59.1(8.5) | 48.2(2.0) | 75.9(3.5) | 31.9(1.3) | 31.8(4.2) | 26.3(2.2) | 38.2(3.2) |
| | LDS | 60.1(2.4) | 60.4(6.2) | 47.0(1.3) | 71.3(2.5) | 31.5(2.0) | 33.2(3.5) | 24.4(3.0) | 38.0(2.4) |
| | INFOGRAPH | 54.9(1.7) | 46.8(1.0) | 43.0(1.9) | 72.3(3.2) | 29.3(1.8) | 27.3(1.4) | 22.6(1.2) | 39.2(4.3) |
| | GREA | 60.3(1.9) | 49.0(4.4) | 48.1(2.5) | 80.7(4.2) | 32.3(1.6) | 26.7(2.7) | 27.2(2.3) | 44.7(6.1) |
| | SGIR | **53.0(0.5)** | **45.4(2.7)** | **42.5(2.8)** | **68.6(2.6)** | **26.6(0.4)** | **24.0(2.2)** | 23.0(1.3) | **33.4(3.0)** |
| O$_2$Perm | GNN | 183.5(33.4) | 6.3(3.2) | 14.6(6.6) | 464.0(85.3) | 7.0(1.8) | 2.4(0.7) | 3.9(1.1) | 29.9(7.2) |
| | RANKSIM | 165.7(27.4) | 3.9(1.4) | 13.0(2.0) | 420.7(69.7) | 5.9(1.4) | **1.8(0.3)** | 3.6(1.7) | 26.6(6.7) |
| | BMSE | 190.4(33.4) | 26.4(21.6) | 27.0(16.4) | 454.3(88.9) | 25.7(14.8) | 14.9(11.7) | 15.9(9.6) | 63.2(23.5) |
| | LDS | 180.0(23.0) | 6.6(4.0) | 11.8(2.0) | 456.3(60.2) | 7.6(1.6) | 2.4(0.6) | 4.7(1.4) | 33.6(9.2) |
| | INFOGRAPH | 199.5(31.5) | 7.5(7.2) | 13.0(1.8) | 505.5(78.2) | 7.8(1.9) | 2.3(0.5) | 5.1(2.2) | 34.8(8.5) |
| | GREA | 182.5(30.0) | 9.0(8.5) | 14.4(4.9) | 458.8(79.2) | 7.1(1.3) | 2.1(0.5) | 4.4(1.3) | 31.7(5.0) |
| | SGIR | **150.9(17.8)** | **3.8(1.1)** | 12.2(0.6) | **382.8(46.9)** | **5.8(0.4)** | 2.1(0.7) | **3.3(0.8)** | **24.4(6.8)** |

The best mean is **bolded**. The best baseline is underlined

labels and augments for data examples on all the possible label ranges, whereas all the baseline models suffer from the bias caused by imbalanced annotations.

**Effectiveness in few-shot label ranges**: The performance improvements of SGIR on graph regression tasks are simultaneously from three different label ranges: *many-shot region*, *medium-shot region*, and *few-shot region*. By looking at the results of baselines, we find that the best performance at a particular range would sacrifice the performance at a different label range. For example, on the Mol-Lipo and Mol-FreeSolv datasets, while GREA is the second best and best baseline, respectively, in the *many-shot region*, its performance in the *few-shot region* is worse than the basic GNN models. Similarly, on the Mol-FreeSolv dataset, LDS reduces the MAE from GNN relatively by +3.5% in the *few-shot region* with a trade-off of a -29% performance decrease in the *many-shot region*. Compared to baselines, the improvements from SGIR in the under-represented label ranges are theoretically guaranteed without sacrificing the performance in the well-represented label range. And the experimental observations support the theoretical guarantee, even in more challenging scenarios, *i.e.*, predictions in the label ranges of fewer training shots on smaller datasets. Specifically, SGIR reduces MAE relatively by 30.3% and 9.0% in the *few-shot region* on Mol-FreeSolv and $O_2$Perm. Because SGIR leverages the mutual enhancement of model construction and data balancing: the gradually balanced training data reduce model bias to popular labels; the less biased model improves the quality of pseudo-labels and augmented examples in the *few-shot region*.

**Effectiveness on different graph regression tasks**: We observe that the improvements on molecule regression tasks are more significant than those on polymer regression tasks. We hypothesize the reasons to be (1) the quality of unlabeled source data and (2) the size of the label space. First, unlabeled graphs consist of more than a hundred thousand unlabeled small molecule graphs from QM9 (Ramakrishnan et al. 2014) and around ten thousand polymers (macromolecules) from Liu et al. (2022). The massive quantity of unlabeled molecules make it easier to have good quality pseudo-labels and augmented examples for the three small molecule regression tasks on Mol-Lipo, Mol-ESOL, and Mol-FreeSolv (Ramakrishnan et al. 2014). Because the majority of unlabeled molecule graphs have a big domain gap with the polymer regression tasks, the quality of expanded training data in polymer regression tasks would be relatively worse than the quality of those in molecule regression. This inspires us to collect more polymer data in the future, even if their properties could not be annotated. Second, Fig. 5.2 has shown that the label ranges in the polymer regression tasks are usually much wider than the ranges in the molecule regression tasks. This poses a great challenge for accurate predictions, especially when we train with a small dataset.

**Effectiveness on age prediction**: Besides molecules and polymers, Table 5.3 shows more results by comparing different methods on the Superpixel-Age dataset. SGIR consistently improves the model performance compared to the best baselines in different label ranges. In the entire label range, SGIR reduces the MAE (GM) relatively by +4.7% (+3.6%). The advantages mainly stem from the enhancements in the *few-shot region*, as demonstrated

## 5.5 Experiments

**Table 5.3** Results of MEAN(STD) on the age prediction using graphs from image superpixels

| | MAE ↓ | | | | GM ↓ | | | |
|---|---|---|---|---|---|---|---|---|
| | All | Many-shot | Med.-shot | Few-shot | All | Many-shot | Med.-shot | Few-shot |
| GNN | 14.583(0.413) | 10.524(0.994) | 11.698(0.404) | 22.127(0.780) | 9.996(0.386) | 7.265(0.858) | 7.910(0.492) | 18.404(0.673) |
| RANKSIM | 14.464(0.401) | 10.468(0.759) | 11.610(0.774) | 21.910(0.700) | 9.606(0.303) | 6.936(0.598) | 7.721(0.660) | 17.534(1.768) |
| BMSE | 15.179(0.594) | 10.639(2.303) | 12.201(0.900) | 23.321(2.525) | 10.419(0.393) | 7.249(1.526) | 8.659(0.827) | 19.719(4.318) |
| LDS | 14.674(0.191) | 10.972(0.495) | 11.985(0.627) | 21.623(0.926) | 9.867(0.291) | 7.317(0.672) | 7.997(0.633) | 17.298(0.957) |
| INFOGRAPH | 14.515(0.605) | 10.610(1.063) | 11.150(0.158) | 22.476(1.147) | 9.879(0.524) | 7.391(0.995) | 7.377(0.333) | 18.969(1.873) |
| GREA | 14.682(0.300) | 10.283(0.503) | 11.999(0.585) | 22.329(0.570) | 10.037(0.438) | 7.051(0.455) | 8.273(0.565) | 18.142(1.276) |
| SGIR | **13.787**(0.123) | **10.171**(0.4156) | **11.066**(0.389) | **20.687**(0.839) | **9.261**(0.221) | **6.928**(0.355) | **7.247**(0.593) | **16.769**(1.418) |

The best mean is **bold**. The best baseline is underlined

in Table 5.3, which shows an improvement of +4.3% and +3.1% on the MAE and GM metrics, respectively. Different from LDS, SGIR improves the model performance for the under-represented and well-represented label ranges at the same time. Table 5.3 showcases that the empirical advantages of SGIR could generalize across different domains.

### 5.5.3 RQ2: Ablation Studies and Sensitivity Analysis

Four ablation studies and one sensitivity analysis are (1) $\mathcal{G}_{conf}$ and $\mathcal{H}_{aug}$ for data balancing; (2) mutually enhanced iterative process; (3) choices of confidence score; (4) quality and diversity of the label-anchored mixup; and (5) The sensitivity analysis for the label interval number $C$.

**Effect of balancing data with different components in $\mathcal{G}_{conf}$ and $\mathcal{H}_{aug}$.** Studies on molecule regression tasks in Table 5.4 present how SGIR improves the initial supervised performance to the most advanced semi-supervised performance step by step. In the first

**Table 5.4** Ablation study on molecular regression with the metric MAE (↓)

|  |  | $\sigma$ | $p$ | $(\tilde{\mathbf{h}}, \tilde{y})$ | All | Many-shot | Med.-shot | Few-shot |
|---|---|---|---|---|---|---|---|---|
| Mol-Lipo | w/o $\mathcal{G}_{unlbl}$ | | | | 0.477(0.014) | 0.378(0.030) | 0.440(0.011) | 0.600(0.006) |
| | | ✓ | ✗ | ✗ | 0.448(0.006) | 0.371(0.004) | 0.421(0.012) | 0.543(0.016) |
| | | ✗ | ✓ | ✗ | 0.446(0.008) | **0.356**(0.003) | **0.407**(0.011) | 0.564(0.016) |
| | | ✓ | ✓ | ✗ | 0.442(0.012) | 0.372(0.007) | 0.415(0.004) | 0.533(0.026) |
| | | ✗ | ✗ | ✓ | 0.456(0.007) | 0.372(0.014) | 0.436(0.010) | 0.549(0.005) |
| | | ✓ | ✓ | ✓ | **0.432**(0.012) | 0.357(0.019) | 0.413(0.017) | **0.515**(0.020) |
| Mol-ESOL | w/o $\mathcal{G}_{unlbl}$ | | | | 0.477(0.027) | 0.375(0.014) | 0.432(0.042) | 0.637(0.042) |
| | | ✓ | ✗ | ✗ | 0.475(0.014) | 0.369(0.014) | 0.446(0.017) | 0.618(0.039) |
| | | ✗ | ✓ | ✗ | 0.480(0.017) | 0.380(0.035) | 0.440(0.017) | 0.630(0.020) |
| | | ✓ | ✓ | ✗ | 0.468(0.007) | 0.379(0.012) | 0.425(0.013) | 0.612(0.028) |
| | | ✗ | ✗ | ✓ | 0.474(0.010) | **0.353**(0.018) | 0.450(0.009) | 0.623(0.027) |
| | | ✓ | ✓ | ✓ | **0.457**(0.015) | 0.370(0.022) | **0.411**(0.011) | **0.604**(0.024) |
| Mol-FreeSolv | w/o $\mathcal{G}_{unlbl}$ | | | | 0.619(0.019) | **0.525**(0.022) | 0.590(0.035) | 1.000(0.072) |
| | | ✓ | ✗ | ✗ | 0.604(0.020) | 0.557(0.037) | 0.560(0.029) | 0.903(0.055) |
| | | ✗ | ✓ | ✗ | 0.660(0.028) | 0.574(0.015) | 0.650(0.036) | 0.941(0.066) |
| | | ✓ | ✓ | ✗ | 0.568(0.029) | 0.538(0.020) | **0.520**(0.045) | 0.831(0.132) |
| | | ✗ | ✗ | ✓ | 0.593(0.045) | 0.536(0.033) | 0.542(0.067) | 0.947(0.062) |
| | | ✓ | ✓ | ✓ | **0.563**(0.026) | 0.535(0.038) | 0.528(0.046) | **0.777**(0.061) |

$\sigma$ is the confidence score. $p$ is the reverse sampling. $(\tilde{\mathbf{h}}, \tilde{y})$ is the label-anchored mixup

line for each dataset, we use only imbalanced training data $\mathcal{G}_{\text{conf}}$ to train the regression model and observe that the model performs badly in the *few-shot region*. The fourth line for each dataset combines the use of regression confidence $\sigma$ and the reverse sampling $p$ to produce $\mathcal{G}_{\text{conf}}$. It improves the MAE performance in the *few-shot region* relatively by +11.2%, +3.2%, and +15.9% on the Mol-Lipo, Mol-ESOL, and Mol-FreeSolv datasets, respectively. The label-anchored mixup algorithm produces the augmented graph representations $\mathcal{H}_{\text{aug}}$ for the under-represented label ranges. By applying $\mathcal{H}_{\text{aug}}$ with $\mathcal{G}_{\text{conf}}$, the last line continues improving the MAE performance in the *few-shot region* (compared to the third line) relatively by +3.3%, +1.3%, and +6.5% on the Mol-Lipo, Mol-ESOL, and Mol-FreeSolv datasets, respectively. Because the use of $\mathcal{H}_{\text{aug}}$ provides a chance to lead the label distributions of training data closer to a perfect balance. Specifically, the effect of semi-supervised pseudo-labeling, or $\mathcal{G}_{\text{conf}}$, comes from the regression confidence $\sigma$ and reverse sampling rate $p$. Results on Mol-ESOL and Mol-FreeSolv show that without the confidence $\sigma$ (the second line), reverse sampling was useless due to heavy label noise. Results on all molecule datasets indicate that without the reverse sampling rate $p$ (the first line), the improvement to *few-shot region* by pseudo-labels was limited.

**Effect of iterative self-training.** Figure 5.4 confirms that model learning and balanced training data mutually enhance each other in SGIR. Because we find that the model performance gradually approximates and outperforms the best baseline in the entire label range. It also indicates that the quality of the training data is steadily improved over iterations.

**Effect of regression confidence measurements.** Table 5.5 shows that compared to existing methods that could define regression confidence, the measurement we define and use, GRATION, is the best option for evaluating the quality of pseudo-labels in graph regression tasks. Because GRATION uses various environments subgraphs, which provide diverse perturbations for robust graph learning (Liu et al. 2022). We also observe that DROPOUT can be a good alternative of GRATION. DROPOUT has extensive assessments (Gal and Ghahramani 2016) and makes it possible for SGIR to be extended to regression tasks for other data modalities such as images and texts.

**Effect of label-anchored mixup augmentation.** We implement $z_i$ using $\mathcal{G}_{\text{imb}}$ to improve the augmentation quality and $\mathcal{G}_{\text{imb}} \cup \mathcal{G}_{\text{unlbl}}$ to improve the diversity. Table 5.6 presents empirical studies to support the idea. It shows that when many noisy representation vectors from unlabeled graphs are included in the interval center $z_i$, the quality of augmented examples is relatively low, which degrades the model performance in different label ranges. On the other hand, the representations of unlabeled graphs improve the diversity of the augmented examples when we assign low mixup weights to them as in Eq. (5.5). Considering both quality and diversity, the effectiveness of the algorithm is further demonstrated in Table 5.4 by significantly reducing the errors for rare labels. From the fifth line of each dataset in Table 5.4, we find that it is also promising to directly use the label-anchored mixup augmentation (as $\mathcal{G}_{\text{imb}} \cup \mathcal{H}_{\text{aug}}$) for data balancing. Although its performance may be inferior to the performance using $\mathcal{G}_{\text{imb}} \cup \mathcal{G}_{\text{conf}}$ (as the third line of each dataset in Table 5.4), the label-anchored

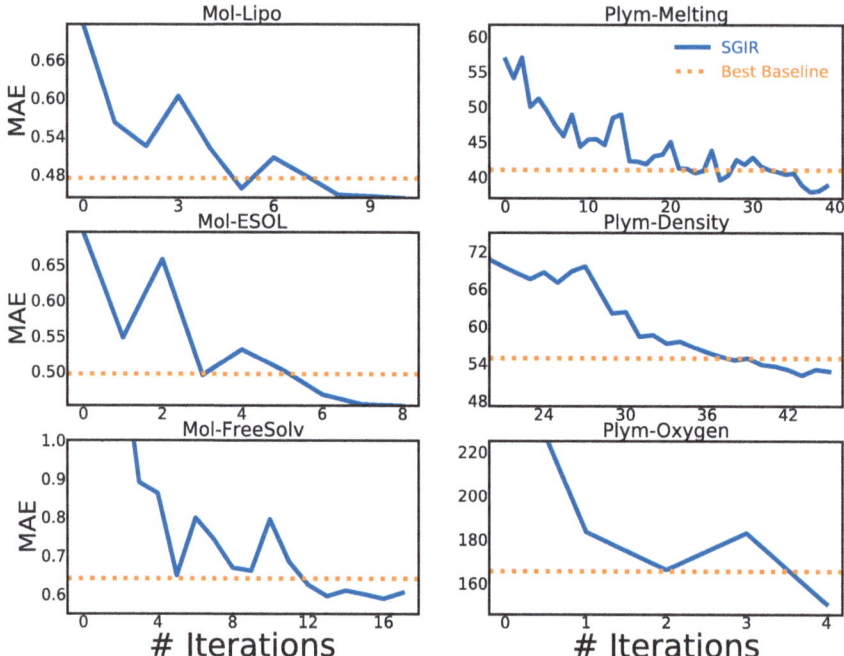

**Fig. 5.4** SGIR test performance (entire label range) through multiple self-training iterations. MAE for PolyDensity is scaled by ×1, 000. The iterative self-training algorithm is effective for gradually improving the quality of training data

mixup algorithm could be improved by the quality of the augmented examples to close the gap with real molecular graphs.

**Sensitivity analysis for the label interval number $C$.** We find the best values of $C$ in main experiments using the validation set for pseudo-labeling and label-anchored mixup. We suggest setting the number $C$ to approximately 100 for pseudo-labeling and around 1,000 for label-anchored mixup. Specifically, sensitivity analysis is conducted on the $O_2$Perm dataset to analyze the effect of the number $C$. Results are presented in Fig. 5.5. We observe that SGIR is robust to a wide range of choices for the number of intervals.

## 5.6 Conclusion

In this chapter, we introduced a self-training framework, called SGIR, to gradually reduce the model bias of data imbalance through multiple iterations. In each iteration, SGIR selected more high-quality pseudo-labels for rare label values and continued augmenting training data to approximate the perfectly balanced label distribution. Experiments demonstrated the effectiveness and reasonable design of the framework, especially in materials science.

## 5.6 Conclusion

**Table 5.5** Investigation on choices of regression confidence with the metric MAE ($\downarrow$)

|  | Choice of $\sigma$ | All | Many-shot | Med.-shot | Few-shot |
|---|---|---|---|---|---|
| Mol-Lipo | SIMPLE | 0.481(0.010) | 0.389(0.007) | 0.440(0.013) | 0.603(0.023) |
|  | DROPOUT | 0.450(0.026) | **0.365**(0.031) | **0.420**(0.022) | 0.555(0.037) |
|  | CERTI | 0.452(0.011) | 0.384(0.018) | 0.433(0.013) | **0.532**(0.010) |
|  | DER | 1.026(0.033) | 0.604(0.035) | 0.760(0.016) | 1.672(0.111) |
|  | GRATION | **0.448**(0.006) | 0.371(0.004) | 0.421(0.012) | 0.543(0.016) |
| Mol-ESOL | SIMPLE | 0.499(0.016) | 0.397(0.023) | 0.457(0.018) | 0.656(0.033) |
|  | DROPOUT | 0.483(0.011) | 0.381(0.027) | **0.443**(0.018) | 0.636(0.027) |
|  | CERTI | 0.487(0.030) | 0.389(0.039) | 0.439(0.024) | 0.647(0.043) |
|  | DER | 0.918(0.135) | 0.776(0.086) | 0.826(0.098) | 1.182(0.245) |
|  | GRATION | **0.475**(0.014) | **0.369**(0.014) | 0.446(0.017) | **0.618**(0.039) |
| Mol-FreeSolv | SIMPLE | 0.697(0.056) | 0.616(0.025) | 0.663(0.033) | 1.054(0.260) |
|  | DROPOUT | 0.639(0.013) | 0.578(0.060) | 0.589(0.017) | 1.005(0.140) |
|  | CERTI | 0.654(0.049) | 0.589(0.046) | 0.611(0.053) | 0.999(0.130) |
|  | DER | 1.483(0.174) | 1.180(0.162) | 1.450(0.188) | 2.480(0.373) |
|  | GRATION | **0.604**(0.020) | **0.557**(0.037) | **0.560**(0.029) | **0.903**(0.055) |

We disable all other SGIR components except the regression confidence score. Our confidence score (**GRation**) in Eq. (5.1) removes noise more effectively than others in graph regression tasks

**Table 5.6** Nine options on the implementation of the label-anchored mixup in Eq. (5.5)

| Additional source | | $O_2$Perm | | | |
|---|---|---|---|---|---|
| $z_i$ | $h_j$ | All | Many-shot | Med.-shot | Few-shot |
| None | None | 165.5(12.2) | 4.7(1.7) | 16.5(7.2) | 417.4(31.1) |
| None | $\mathcal{G}_{\text{conf}}$ | 158.1(17.0) | 4.1(0.7) | **11.3**(0.7) | 401.9(45.1) |
| None | $\mathcal{G}_{\text{unlbl}}$ | **150.9**(17.8) | **3.8**(1.1) | 12.2(0.6) | **382.8**(46.9) |
| $\mathcal{G}_{\text{conf}}$ | None | 166.0(18.2) | 11.9(11.3) | 12.6(0.9) | 414.0(52.6) |
| $\mathcal{G}_{\text{conf}}$ | $\mathcal{G}_{\text{conf}}$ | 158.8(8.4) | 7.7(8.9) | 15.4(7.8) | 397.5(15.4) |
| $\mathcal{G}_{\text{conf}}$ | $\mathcal{G}_{\text{unlbl}}$ | 169.5(56.1) | 4.5(1.2) | 12.7(1.8) | 430.4(145.0) |
| $\mathcal{G}_{\text{unlbl}}$ | None | 173.1(30.3) | 3.7(0.4) | 13.5(1.4) | 440.0(79.3) |
| $\mathcal{G}_{\text{unlbl}}$ | $\mathcal{G}_{\text{conf}}$ | 174.5(9.3) | 8.1(3.3) | 11.9(0.9) | 440.4(25.5) |
| $\mathcal{G}_{\text{unlbl}}$ | $\mathcal{G}_{\text{unlbl}}$ | 156.3(20.5) | 8.2(2.9) | 12.9(0.9) | 392.3(50.6) |

Except for the imbalanced labeled graphs $\mathcal{G}_{\text{imb}}$, the additional source of the interval representation $z_i$ and the real graph representation $h_j$ could be $\mathcal{G}_{\text{conf}}$ or $\mathcal{G}_{\text{unlbl}}$. We find that source $z_i$ from $\mathcal{G}_{\text{imb}}$ and source $h_j$ from $\mathcal{G}_{\text{unlbl}}$ are usually the best

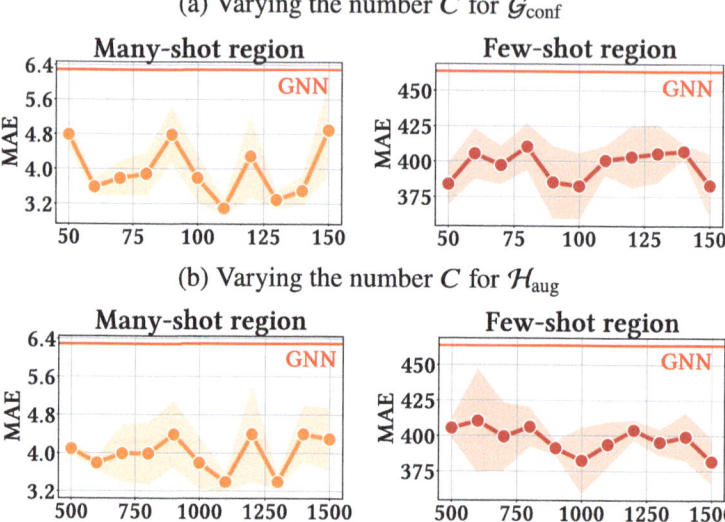

**Fig. 5.5** Sensitivity analysis on the number of label intervals ($C$) for pseudo-labeling selection $\mathcal{G}_{\text{conf}}$ (top) and label-anchored mixup algorithm $\mathcal{H}_{\text{aug}}$ (bottom). Results are drawn on the O$_2$Perm

## References

R. Achanta, A. Shaji, K. Smith, A. Lucchi, P. Fua, and S. Süsstrunk. Slic superpixels compared to state-of-the-art superpixel methods. *IEEE transactions on pattern analysis and machine intelligence*, 34(11):2274–2282, 2012.

A. Amini, W. Schwarting, A. Soleimany, and D. Rus. Deep evidential regression. *Advances in Neural Information Processing Systems*, 33:14927–14937, 2020.

P. L. Bartlett and S. Mendelson. Rademacher and gaussian complexities: Risk bounds and structural results. *Journal of Machine Learning Research*, 3(Nov):463–482, 2002.

D. Berthelot, R. Roelofs, K. Sohn, N. Carlini, and A. Kurakin. Adamatch: A unified approach to semi-supervised learning and domain adaptation. *International Conference on Learning Representations*, 2022.

K. Cao, C. Wei, A. Gaidon, N. Arechiga, and T. Ma. Learning imbalanced datasets with label-distribution-aware margin loss. *Advances in neural information processing systems*, 32, 2019.

L. Carratino, M. Cissé, R. Jenatton, and J.-P. Vert. On mixup regularization. *arXiv preprint* arXiv:2006.06049, 2020.

C. Cortes, M. Mohri, M. Riley, and A. Rostamizadeh. Sample selection bias correction theory. In *Algorithmic Learning Theory: 19th International Conference, ALT 2008, Budapest, Hungary, October 13-16, 2008. Proceedings 19*, pages 38–53. Springer, 2008.

Y. Gal and Z. Ghahramani. Dropout as a bayesian approximation: Representing model uncertainty in deep learning. In *international conference on machine learning*, pages 1050–1059. PMLR, 2016.

Y. Gong, G. Mori, and F. Tung. Ranksim: Ranking similarity regularization for deep imbalanced regression. *International Conference on Machine Learning*, 2022.

S. M. Kakade, K. Sridharan, and A. Tewari. On the complexity of linear prediction: Risk bounds, margin bounds, and regularization. *Advances in neural information processing systems*, 21, 2008.

B. Knyazev, G. W. Taylor, and M. Amer. Understanding attention and generalization in graph neural networks. *Advances in neural information processing systems*, 32, 2019.

G. Liu, T. Zhao, J. Xu, T. Luo, and M. Jiang. Graph rationalization with environment-based augmentations. In *Proceedings of the 28th ACM SIGKDD Conference on Knowledge Discovery and Data Mining*, pages 1069–1078, 2022.

Z. Liu, Z. Wang, H. Guo, and Y. Mao. Over-training with mixup may hurt generalization. In *The Eleventh International Conference on Learning Representations*, 2023. URL https://openreview.net/forum?id=JmkjrlVE-DG.

R. Ma and T. Luo. Pi1m: a benchmark database for polymer informatics. *Journal of Chemical Information and Modeling*, 60(10):4684–4690, 2020.

G. J. McLachlan. Iterative reclassification procedure for constructing an asymptotically optimal rule of allocation in discriminant analysis. *Journal of the American Statistical Association*, 70(350):365–369, 1975.

D. Mendez, A. Gaulton, A. P. Bento, J. Chambers, M. De Veij, E. Félix, M. P. Magariños, J. F. Mosquera, P. Mutowo, M. Nowotka, et al. Chembl: towards direct deposition of bioassay data. *Nucleic acids research*, 47(D1):D930–D940, 2019.

S. Moschoglou, A. Papaioannou, C. Sagonas, J. Deng, I. Kotsia, and S. Zafeiriou. Agedb: the first manually collected, in-the-wild age database. In *proceedings of the IEEE conference on computer vision and pattern recognition workshops*, pages 51–59, 2017.

S. Otsuka, I. Kuwajima, J. Hosoya, Y. Xu, and M. Yamazaki. Polyinfo: Polymer database for polymeric materials design. In *2011 International Conference on Emerging Intelligent Data and Web Technologies*, pages 22–29. IEEE, 2011.

R. Ramakrishnan, P. O. Dral, M. Rupp, and O. A. Von Lilienfeld. Quantum chemistry structures and properties of 134 kilo molecules. *Scientific data*, 1(1):1–7, 2014.

J. Ren, M. Zhang, C. Yu, and Z. Liu. Balanced mse for imbalanced visual regression. In *Proceedings of the IEEE/CVF Conference on Computer Vision and Pattern Recognition*, 2022.

M. Sugiyama and A. J. Storkey. Mixture regression for covariate shift. *Advances in neural information processing systems*, 19, 2006.

B. Sun, J. Feng, and K. Saenko. Return of frustratingly easy domain adaptation. In *Proceedings of the AAAI conference on artificial intelligence*, volume 30, 2016.

F.-Y. Sun, J. Hoffman, V. Verma, and J. Tang. Infograph: Unsupervised and semi-supervised graph-level representation learning via mutual information maximization. In *International Conference on Learning Representations*, 2020.

N. Tagasovska and D. Lopez-Paz. Single-model uncertainties for deep learning. *Advances in Neural Information Processing Systems*, 32, 2019.

A. Thornton, L. Robeson, B. Freeman, and D. Uhlmann. Polymer gas separation membrane database, 2012. URL https://research.csiro.au/virtualscreening/membrane-database-polymer-gas-separation-membranes/.

J. Tian, Y.-C. Liu, N. Glaser, Y.-C. Hsu, and Z. Kira. Posterior re-calibration for imbalanced datasets. *Advances in Neural Information Processing Systems*, 33:8101–8113, 2020.

V. Verma, A. Lamb, C. Beckham, A. Najafi, I. Mitliagkas, D. Lopez-Paz, and Y. Bengio. Manifold mixup: Better representations by interpolating hidden states. In *International Conference on Machine Learning*, pages 6438–6447. PMLR, 2019.

Y. Wang, W. Wang, Y. Liang, Y. Cai, and B. Hooi. Mixup for node and graph classification. In *Proceedings of the Web Conference 2021*, pages 3663–3674, 2021.

C. Wei, K. Sohn, C. Mellina, A. Yuille, and F. Yang. Crest: A class-rebalancing self-training framework for imbalanced semi-supervised learning. In *Proceedings of the IEEE/CVF Conference on Computer Vision and Pattern Recognition*, pages 10857–10866, 2021.

Y.-X. Wu, X. Wang, A. Zhang, X. He, and T. seng Chua. Discovering invariant rationales for graph neural networks. In *ICLR*, 2022.

Z. Wu, B. Ramsundar, E. N. Feinberg, J. Gomes, C. Geniesse, A. S. Pappu, K. Leswing, and V. Pande. Moleculenet: a benchmark for molecular machine learning. *Chemical science*, 9(2):513–530, 2018.

Q. Xie, M.-T. Luong, E. Hovy, and Q. V. Le. Self-training with noisy student improves imagenet classification. In *Proceedings of the IEEE/CVF conference on computer vision and pattern recognition*, pages 10687–10698, 2020.

K. Xu, W. Hu, J. Leskovec, and S. Jegelka. How powerful are graph neural networks? In *International Conference on Learning Representations*, 2019. URL https://openreview.net/forum?id=ryGs6iA5Km.

Y. Yang, K. Zha, Y. Chen, H. Wang, and D. Katabi. Delving into deep imbalanced regression. In *International Conference on Machine Learning*, pages 11842–11851. PMLR, 2021.

H. Yao, Y. Wang, L. Zhang, J. Y. Zou, and C. Finn. C-mixup: Improving generalization in regression. *Advances in Neural Information Processing Systems*, 35:3361–3376, 2022.

Q. Yuan, M. Longo, A. W. Thornton, N. B. McKeown, B. Comesana-Gandara, J. C. Jansen, and K. E. Jelfs. Imputation of missing gas permeability data for polymer membranes using machine learning. *Journal of membrane science*, 627:119207, 2021.

L. Zhang, Z. Deng, K. Kawaguchi, A. Ghorbani, and J. Zou. How does mixup help with robustness and generalization? In *International Conference on Learning Representations*, 2021. URL https://openreview.net/forum?id=8yKEo06dKNo.

Y. Zhang, B. Kang, B. Hooi, S. Yan, and J. Feng. Deep long-tailed learning: A survey. *IEEE Transactions on Pattern Analysis and Machine Intelligence*, 2023.

T. Zhao, T. Jiang, N. Shah, and M. Jiang. A synergistic approach for graph anomaly detection with pattern mining and feature learning. *IEEE Transactions on Neural Networks and Learning Systems*, 33(6):2393–2405, 2021.

# 6 Generative Modeling: Data-Centric Learning from Unlabeled Graphs with Diffusion Model

## 6.1 Introduction

Graphs such as molecules and polymers are found to have attractive properties in drug and material discovery (Böhm et al. 2004; Huang et al. 2021), but annotating them requires specialized knowledge, as well as lengthy and costly experiments in wet labs (Cormack and Elorza 2004). It is important for graph property predictors to learn *useful knowledge* from unlabeled graphs.

Self-supervised learning (Hu et al. 2019; Rong et al. 2020; You et al. 2021; Kim et al. 2022) utilizes unlabeled graphs to learn through *predictive tasks* or *contrastive tasks* to represent and transfer the knowledge as *model parameters*. Despite the empirical success in language and vision (Brown et al. 2020; He et al. 2022), their performance on graph data applications remains unsatisfactory because of the significant gap between the graph self-supervised task and the graph label prediction task. Models trained on node attribute prediction (Hu et al. 2019) as a simple *predictive* self-supervised task extract too limited knowledge from the graph structure, which has been observed after too fast convergence (Sun et al. 2022). More complex tasks like graph context prediction (Hu et al. 2019; Zhang et al. 2021) may transfer knowledge that conflicts with downstream tasks. Aromatic rings, for instance, are a prevalent structure in molecules (Maziarka et al. 2020) and are considered valuable in context prediction tasks (Zhang et al. 2021). However, graph properties such as oxygen permeability can be more related to non-aromatic rings in some cases (Liu et al. 2022), which is overlooked if not using tailored predictive tasks specifically for downstream tasks. As predictive tasks strive for universality, the transferred knowledge may force models to focus more on aromatic rings, leading to poor prediction.

On the other line, *contrastive* tasks (You et al. 2021; Kim et al. 2022) aim to learn the similarity between original and perturbed graphs. However, the learned similarity can hardly generalize across tasks (Kim et al. 2022). First, perturbations without domain knowledge, *e.g.*, bioisosteres, do not preserve broad biological properties (Sun et al. 2021). Second, it is difficult, if not impossible, to find universal perturbations that generalize to diverse property prediction tasks. For example, bioisosteric (subgraph) replacements produce similar biological properties for molecules. And they may reduce toxicity (Brown 2014). So, contrastive tasks with bioisosteric replacement enforce the similarity between toxic and non-toxic molecules. However, models pre-trained on such contrastive tasks hurt the performance on downstream tasks, *e.g.*, toxicity prediction.

The *data-centric* idea avoids the use of self-supervised tasks that are not appropriate. We use a diffusion probabilistic model (known as *diffusion model*) to capture the data distribution of *unlabeled graphs*, leveraging its capability of distribution coverage, stationarity, and scalability (Dhariwal and Nichol 2021). At the stage of performing a particular property prediction task, the reverse process, guided by novel task-related optimization objectives, generates new task-specific labeled examples. Minimal sufficient knowledge from the unlabeled data is transferred into these examples, instead of uninterpretable model parameters, and then to enhance the training of prediction models.

To implement the idea, we present the *Data-Centric Transfer* framework (DCT) based on a diffusion model for graph data, as shown in Fig. 6.1b. It aims to transfer minimal sufficient knowledge from unlabeled graphs to property predictors by data augmentation. The diffusion model gradually adds Gaussian noise to a graph from which a score function (*i.e.*, the gradient of the log probability density) is then learned to estimate the noise step by step to reverse the process. DCT trains the diffusion model on the unlabeled graphs to get ready to augment any labeled dataset. Given a labeled graph from a particular task (*i.e.*, type of property), the diffusion model adds noise to perturb it by a few steps and then generates a new graph through the score function. The new graph could be close to the distribution of the unlabeled graphs for diversity, however, it would lose the relatedness to the target task. So, we add two task-related objectives into the score function to guide the reverse process. When a predictor model $f$ has been trained on the task, given an original labeled graph $G$, the first objective is to optimize the new graph $G'$ to *sufficiently* preserve the label of $G$ with $f$. The second objective is to optimize $G'$ to be very different from (*i.e.*, minimally similar to) $G$. These two objectives ensure that $G'$ carries minimal sufficient knowledge from the unlabeled graphs to be an augmentation of $G$. DCT iteratively generates new examples to augment the labeled dataset and progressively trains the prediction model with it.

We illustrate the relationship between DCT and various data-centric methods, such as graph data augmentation and self-training, in Fig. 6.2. Perturbing edges, deleting nodes, and masking attributes (Rong et al. 2019; Trivedi et al. 2022) are heuristic methods for graph data augmentation. The augmented knowledge from them is mainly controlled by human prior knowledge on the perturbations and it often fails to be close to the task, *i.e.*, random perturbations hardly preserve labels for the augmented graphs. The learning to augment approaches learn from labeled graphs to perturb graph structures (Luo et al. 2022), to estimate graphons for different classes (Han et al. 2022), or to split the latent space for augmentation (Liu et al. 2022). Although these approaches could preserve labels for the

## 6.1 Introduction

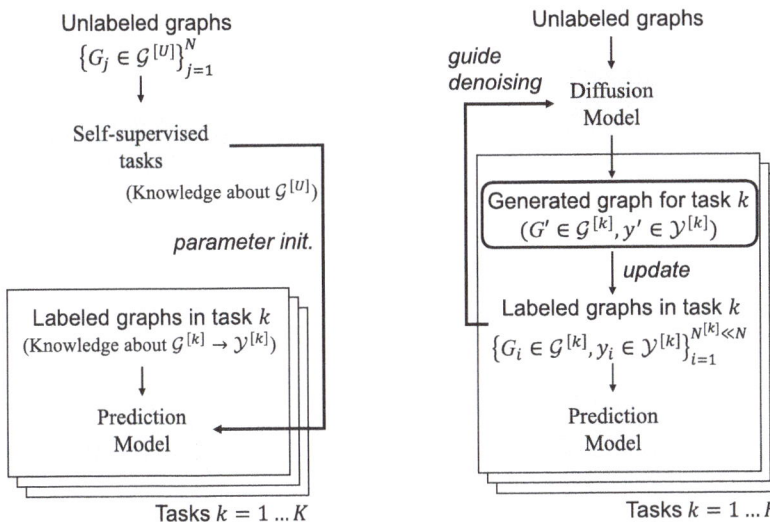

(a) **Existing approach**: Knowledge from self-supervised tasks could not be aligned or even conflict with what predictions need. Parameter initialization could *hardly interpret* how unlabeled graphs were or would be able to improve the models, leading to high prediction errors.

(b) **Data-centric approach**: Target knowledge in labeled graphs guides the denoising process in diffusion model to generate new labeled examples close to the target graph space instead of the unlabeled graph space. The augmented knowledge for the prediction model is *visible* as graphs.

**Fig. 6.1** Comparing the diagrams of the existing approach and the DCT approach to learning from unlabeled graphs for a variety of molecular graph tasks

**Fig. 6.2** Qualitative relationship of graphs from different data-centric approaches on the task relatedness and contained knowledge

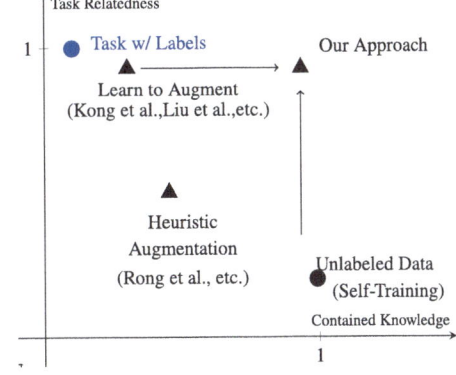

augmented graphs, they introduce less extra knowledge to improve the model prediction. In summary, graph data augmentation is effective in expanding knowledge for limited labels, but it makes no use of unlabeled graphs. Besides, the diversity and richness of the domain knowledge from augmented graphs are far from that contained in a large number of unlabeled graphs. To learn from unlabeled graphs, data-centric approaches like the self-training is assumed to be useful when the unlabeled and labeled data are from the same source. It is less studied when we have a single unified unlabeled source for different tasks. DCT achieves both task relatedness and diversity utilizing a diffusion model to extract knowledge (as the diffusion and reverse processes) from *all the unlabeled graphs*. DCT represents the knowledge as task-specific labeled examples to augment the target dataset, instead of uninterpretable pre-trained model parameters.

We test DCT on *fifteen* graph property prediction datasets from three fields: chemistry (molecules), material science (polymers), and biology (protein-protein interaction graphs). DCT achieves the best performance over all these tasks. We find that the state-of-the-art self-supervised methods often struggle to transfer knowledge to regression tasks, etc. DCT reduces the mean absolute error relatively by 13.4% and 10.2% compared to the best baseline on the molecule and polymer graph regression tasks, respectively.

## 6.2  Problem Definition

Given $K$ property prediction tasks, there are $N^{[k]}$ labeled graph examples for the $k$-th task. They are $\{(G_i, y_i) \mid G_i \in \mathcal{G}^{[k]}, y_i \in \mathcal{Y}^{[k]}\}_{i=1}^{N^{[k]}}$, where $\mathcal{G}^{[k]}$ is the graph space and $\mathcal{Y}^{[k]}$ is the label space of the task. The prediction model with parameters $\theta$ is defined as $f_\theta^{[k]} : \mathcal{G}^{[k]} \to \mathcal{Y}^{[k]}$. $f_\theta^{[k]}$ consists of a GNN and a multi-layer perceptron (MLP). Without the loss of generality, we consider Graph Isomorphism Networks (GIN) (Xu et al. 2018) to encode graph structures. Given a graph $G = (\mathcal{V}, \mathcal{E}) \in \mathcal{G}^{[k]}$ in the task $k$, GIN updates the representation vector of node $v \in \mathcal{V}$ at $l$-layer:

$$\mathbf{h}_v^l = \text{MLP}^l \left( (1 + \epsilon) \cdot \mathbf{h}_v^{l-1} + \sum_{u \in \mathcal{N}(v)} \mathbf{h}_u^{l-1} \right), \tag{6.1}$$

where $\epsilon$ is a learnable scalar and $u \in \mathcal{N}(v)$ is one of node $v$'s neighbor nodes. After stacking $L$ layers, the READOUT function (*e.g.*, summation) gets the graph representation across all the nodes. The predicted label is:

$$\hat{y} = \text{MLP}\left(\text{READOUT}\left(\{\mathbf{h}_v^L \mid v \in G\}\right)\right). \tag{6.2}$$

$f_\theta^{[k]}$ is hard to be well-trained because collecting graph labels at a large scale is difficult. That is $N^{[k]}$ is typically small.

Fortunately, regardless of the tasks, a large number of **unlabeled graphs** are usually available from the same or similar domains. Self-supervised learning methods (Hu et al. 2019) rely on hand-crafted tasks to extract *knowledge* from the unlabeled examples $\{G_j \in \mathcal{G}^{[U]}, j = 1, \ldots, N\}$ as *pre-trained model parameters* $\theta$. The uninterpretable parameters are transferred to warm up the prediction models $\{f_\theta^{[k]}\}_{k=1}^K$ on the $K$ downstream graph property prediction tasks. However, the gap and even conflict between the self-supervised

## 6.3 Data-Centric Transfer Framework: DCT

tasks and the property prediction tasks lead to suboptimal performance of the prediction models. In the next section, we present the DCT framework that transfers knowledge from the unlabeled graphs.

## 6.3 Data-Centric Transfer Framework: DCT

### 6.3.1 Overview of Developed Framework

The goal of data-centric approaches is to augment training datasets by generating useful labeled data examples. Under that, the goal of the data-centric transfer (DCT) framework is to *transfer* the knowledge from unlabeled data into the data augmentation. Specifically, for each graph-label pair $(G^{[k]} \in \mathcal{G}^{[k]}, y^{[k]} \in \mathcal{Y}^{[k]})$ in the task $k$, the framework is expected to output a new example $G'^{[k]}$ with the label $y'^{[k]}$ such that (1) $y'^{[k]} = y^{[k]}$ and (2) $G'^{[k]}$ and $G^{[k]}$ are from the same graph space $\mathcal{G}^{[k]}$. However, if the graph structures of $G'^{[k]}$ and $G^{[k]}$ were too similar, the augmentation would duplicate the original data examples, become useless, and even cause over-fitting. So, the optimal graph data augmentation should *enrich the training data with good diversity as well as preserve the labels of the original graphs*. To achieve this, DCT utilizes a diffusion probabilistic model to first *learn the data distribution from unlabeled graphs* (Sect. 6.3.2). Then DCT adapts the reverse process in the diffusion model to *generate task-specific labeled graphs for data augmentation* (Sect. 6.3.3). Thus, the augmented graphs will be derived from the distribution of a huge collection of unlabeled data for *diversity*. To *preserve the labels*, DCT controls the reverse process with two task-related optimization objectives to transfer *minimal sufficient knowledge* from the unlabeled data. The first objective minimizes an upper bound of mutual information between the augmented and the original graphs in the graph space. The second objective maximizes the probability of the predicted label of augmented graphs being the same as the label of original graphs. The first is for minimal knowledge transfer, and the second is for sufficient knowledge transfer. DCT integrates the two objectives into the reverse process of the diffusion model to guide the generation of new labeled graphs. DCT iteratively trains the graph property predictor (used in the second objective) and creates the augmented training data. To simplify notations, we remove the task superscript $[k]$ in the following sections.

### 6.3.2 Learning Data Distribution from Unlabeled Graphs

The diffusion process for graphs in Fig. 6.3 applies to graph structure and node features. The diffusion model slowly corrupts unlabeled graphs to a standard normal distribution with noise. For graph generation, the model samples noise from the normal distribution and learns a score function to reverse the perturbed noise. Given an unlabeled graph $G$, we use continuous time $t \in [0, T]$ to index multiple diffusion steps $\{G^{(t)}\}_{t=1}^{T}$ on the graph, such

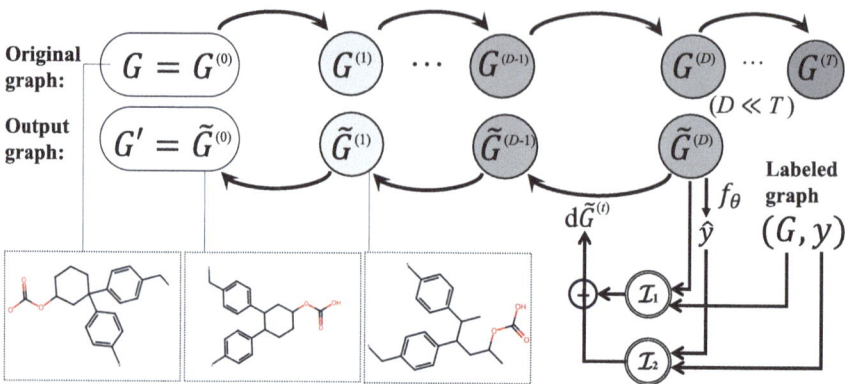

**Fig. 6.3** Diffusion model in DCT: It performs task-specific data augmentation using objectives $\mathcal{I}_1$ and $\mathcal{I}_2$ in the reverse process. The model was trained on unlabeled graphs to learn the general data distribution. Then it generates $(G', y' = y)$ based on $(G, y)$ in the reverse process. It perturbs $G$ with $D$ steps and optimizes $G'$ to be minimally similar to $G$ (Objective $\mathcal{I}_1$) and sufficiently preserve the label of $G$ (Objective $\mathcal{I}_2$)

that $G^{(0)}$ follows the original data distribution and $G^{(T)}$ follows a prior distribution like the normal distribution. The forward diffusion is a stochastic differential equation (SDE) from the graph to the noise:

$$dG^{(t)} = \mathbf{f}\left(G^{(t)}, t\right) dt + g(t) d\mathbf{w}, \quad (6.3)$$

where $\mathbf{w}$ is the standard Wiener process, $\mathbf{f}(\cdot, t) : \mathcal{G} \to \mathcal{G}$ is the drift coefficient and $g(t) : \mathbb{R} \to \mathbb{R}$ is the diffusion coefficient. $\mathbf{f}(G^{(t)}, t)$ and $g(t)$ relate to the amount of noise added to the graph at each infinitesimal step $t$. The reverse-time SDE uses gradient fields or scores of the perturbed graphs $\nabla_{G^{(t)}} \log p_t(G^{(t)})$ for denoising and graph generation from $T$ to 0 (Song et al. 2021):

$$dG^{(t)} = \left[\mathbf{f}(G^{(t)}, t) - g(t)^2 \nabla_{G^{(t)}} \log p_t(G^{(t)})\right] dt + g(t) d\overline{\mathbf{w}}, \quad (6.4)$$

where $p_t(G^{(t)})$ is the marginal distribution at time $t$ in forward diffusion. $\overline{\mathbf{w}}$ is a reverse time standard Wiener process. $dt$ here is an infinitesimal negative time step. The score $\nabla_{G^{(t)}} \log p_t(G^{(t)})$ is unknown in practice and it is approximated by the score function $\mathbf{s}(G^{(t)}, t)$ with score matching techniques (Song et al. 2021). On graphs, Jo et al. (2022) used two GNNs to develop the score function $\mathbf{s}(G^{(t)}, t)$ to de-noise both node features and graph structures.

**Instantiations of SDE on Graphs**: According to Song et al. (2021), we use the Variance Exploding (VE) SDE for the diffusion process. Given the minimal noise $\sigma_{\min}$ and the maximal noise $\sigma_{\max}$, the VE SDE is:

## 6.3 Data-Centric Transfer Framework: DCT

$$dG = \sigma_{\min} \left( \frac{\sigma_{\max}}{\sigma_{\min}} \right)^t \sqrt{2 \log \frac{\sigma_{\max}}{\sigma_{\min}}} d\mathbf{w}, \quad t \in (0, 1] \quad (6.5)$$

The perturbation kernel is derived (Song et al. 2021) as:

$$p_{0t}(G^{(t)} \mid G^{(0)}) = \mathcal{N}\left(G^{(t)}; G^{(0)}, \sigma_{\min}^2 \left(\frac{\sigma_{\max}}{\sigma_{\min}}\right)^{2t} \mathbf{I}\right), \quad t \in (0, 1] \quad (6.6)$$

We follow Jo et al. (2022) to separate the perturbation of adjacency matrix and node features:

$$p_{0t}(G^{(t)} \mid G^{(0)}) = p_{0t}(\mathbf{A}^{(t)} \mid \mathbf{A}^{(0)}) p_{0t}(\mathbf{X}^{(t)} \mid \mathbf{X}^{(0)}). \quad (6.7)$$

### 6.3.3 Generating Task-Specific Labeled Graphs

*Self-training* approaches would develop to either (1) select unlabeled graphs by a graph property predictor or (2) generate graphs directly from the standard diffusion model, and then use the predictor to assign them labels so that the training data could be enriched. However, we have observed that neither of them can guarantee positive impact on the prediction performance. In fact, as shown in Fig. 6.4, they make very little or even negative impact. That is because *the selected or directly-generated graphs are too different from the labeled graph space of the target tasks*. Task details of ten datasets are in Sect. 6.4.1.

Given a labeled graph $(G, y)$ from the original dataset of a specific task, the new labeled graph $(G', y')$ is expected to provide *useful knowledge to augment* the training set. We name it *the augmented graph* throughout this section. The augmented graph is desired to have the following two properties, as in Sect. 6.3.1: **Task relatedness**: As an effective training data point, $G' \in \mathcal{G}$ and $y' \in \mathcal{Y}$ are from the graph/label spaces of the specific task where $(G, y)$ come from and thus transfer sufficient task knowledge into the training set; **Diversity**: If $G'$ was too similar to $G$, the new data point would cause severe over-fitting on the property

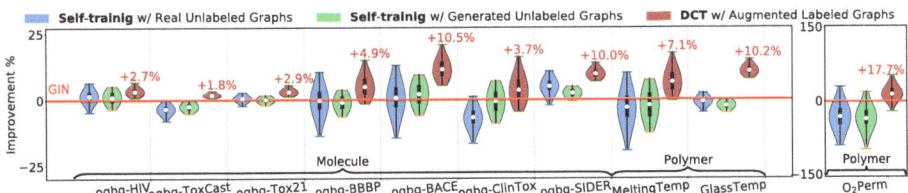

**Fig. 6.4** Relative improvement (increased AUC or reduced MAE) from three data-centric methods (over ten runs), compared to the basic GIN: Blue is for self-training with selected real unlabeled graphs. Green is for self-training with graphs directly generated by a standard diffusion model. Red is for DCT that generates task-specific labeled graphs. The first two often make little or negative impact. DCT has consistent and significant improvement shown as the percentages in red

prediction model. The augmentation aims to learn from unlabeled graphs to create diverse data points, which should contain minimal task knowledge about $G$.

The selected unlabeled graphs used in *self-training* have little task relatedness because the unlabeled data distribution might be too far from the one of the specific task. Existing graph *data augmentation* methods could not create diverse graph examples because they manipulated labeled graphs and did not learn from the unlabeled graphs. The DCT works towards both desired properties by transferring *minimally sufficient knowledge* from the unlabeled graphs: **Sufficiency** is achieved by maximizing the possibility for label preservation (i.e., $y' = y$). It ensures that the knowledge from unlabeled graphs is task-related; **Minimality** refers to the minimization of graph similarity between $G'$ and $G$ to ensure that the augmentation introduces diversity. Both optimizations can be formulated using mutual information $I(\cdot\,;\,\cdot)$ to generate task-specific labeled data $(G', y')$:

**Definition 6.1** *(Sufficiency for Data Augmentation)* The augmented graph $G'$ sufficiently preserves the label of the original graph $G$ if and only if $I(G'; y) = I(G; y)$.

**Definition 6.2** *(Minimal Sufficiency for Data Augmentation)* The Sufficiency is minimal for data augmentation if and only if $I(G'; G) \leq I(\bar{G}; G), \forall \bar{G}$ represents any augmented graph that sufficiently preserves the original graph's label.

Self-supervised tasks applied a similar philosophy in pre-training (Soatto and Chiuso 2016), however, they did not use labeled data from any specific tasks. So the optimizations were unable to extract useful knowledge and transfer it to the downstream (Tian et al. 2020). In DCT that performs task-specific data augmentation, the augmented graphs can be optimized toward the objectives using any labeled graph $G$ and its label $y$:

$$\min_{I_1} \max_{I_2} \; \mathbb{E}_G \left[ I_1\left(G'; G\right) + I_2\left(G'; y\right) \right]. \tag{6.8}$$

For the first objective, we use the leave-one-out variant of InfoNCE (Poole et al. 2019; Oord et al. 2018) as the upper bound estimation. For the $i$th labeled graph $(G_i, y_i)$,

$$I_1 \leq I_{\text{bound}}(G'_i; G_i) = \log \frac{p(G'_i|G_i)}{\sum_{j=1, j\neq i}^{M} p(G'_i|G_j)}, \tag{6.9}$$

where $G'_i$ is the augmented graph. When $G'_i$ is optimized, $G_i$ makes a positive pair; $\{G_j\}$ ($j \neq i$) are $M - 1$ negative samples of labels that do not equal $y_i$. ($M$ is a hyperparameter.) We use cosine similarity and a softmax function to calculate $p(G'_i|G_j) = \frac{\exp(\text{sim}(G'_i, G_j))}{\sum_{j=1}^{M} \exp(\text{sim}(G'_i, G_j))}$. In practice, we extract statistical features of graphs to calculate their similarity. For each molecule and polymer graph, we concatenate the following vectors or values for statistical feature extraction.

## 6.3 Data-Centric Transfer Framework: DCT

- the sum of the degree in the graph;
- the vector indicating the distribution of atom types;
- the vector containing the maximum, minimum and mean values of atoms weights in a molecule or polymer;
- the vector containing the maximum, minimum, and mean values of bond valence.

For each protein-protein interaction ego-graph in the biology field, we use the sorted vector of node degree distribution in the graph as the statistical features.

For the second objective, we denote the predicted label of the augmented graph $G'$ by $f_\theta(G')$. We maximize the log likelihood $\log p\left(y|f_\theta(G')\right)$ to maximize $I_2(G'; y)$. Specifically, after the predictor $f_\theta$ is trained for several epochs on the labeled data, we freeze its parameters and use it to optimize the augmented graphs so they are task-related:

$$\mathcal{L}(G') = I_{\text{bound}}\left(G'; G\right) - \log p\left(y|f_\theta(G')\right). \tag{6.10}$$

**Framework details**: As shown in Fig. 6.3, after the diffusion model learns the data distribution from unlabeled graphs, given a labeled graph $G$ from a specific task, DCT perturbs it for $D$ ($D \ll T$) steps. The perturbed noisy graph, denoted by $\tilde{G}^{(D)}$, stays inside the task-specific graph and label space, rather than the noise distribution (at step $T$). To reverse the noise in it and generate a task-specific augmented example $G'$, DCT integrates the loss function in Eq. (6.10) into the score function $\mathbf{s}(\cdot, t)$ for minimal sufficient knowledge transfer:

$$d\tilde{G}^{(t)} = \left[\mathbf{f}(\tilde{G}^{(t)}, t) - g(t)^2 \left(\mathbf{s}(\tilde{G}^{(t)}, t) - \alpha \nabla_{\tilde{G}^{(t)}} \mathcal{L}(\tilde{G}^{(t)})\right)\right] dt + g(t) d\overline{\mathbf{w}}, \tag{6.11}$$

where $\alpha$ is a scalar for score alignment between $\mathbf{s}$ and $\nabla \mathcal{L}$ to avoid the dominance of any of them: $\alpha = \frac{\|\mathbf{s}(\tilde{G}^{(t)}, t)\|_2}{\|\nabla_{\tilde{G}^{(t)}} \mathcal{L}(\tilde{G}^{(t)})\|_2}$. Because $\tilde{G}^{(t)}$ is an intermediate state in the reverse process, the noise in it may fail the optimizations. So, we design a new sampling method named *double-loop sampling* for accurate loss calculation. It has an inner-loop sampling using Eq. (6.4) to sample $\hat{G}_{(t)}$, as the denoised version of $\tilde{G}^{(t)}$ at the reverse time $t$. Then $\nabla_{\hat{G}} \mathcal{L}(\hat{G}_{(t)})$ is calculated as an alternative for $\nabla_{\tilde{G}^{(t)}} \mathcal{L}(\tilde{G}^{(t)})$. Finally, an outer-loop sampling takes one step to guide denoising using Eq. (6.11).

DCT iteratively creates the augmented graphs $(G', y')$, updates the training dataset $\{(G_i, y_i)\}$, and trains the graph property predictor $f_\theta$. In each iteration, for task $k$, $n \ll N^{[k]}$ labeled graphs of the lowest property prediction loss are selected to create the augmented graphs. The predictor is better fitted to these graphs for more accurate sufficiency estimation of the augmentation.

**Algorithm**: We adapt the Predictor-Corrector (PC) samplers for the graph data augmentation in the reverse process. The algorithm is shown in Algorithm 1. The algorithm of the whole data-centric knowledge transfer framework is presented in Algorithms 2 and 3. In detail, Algorithm 2 corresponds to Sect. 6.3.2 and Algorithm 3 corresponds to Sect. 6.3.3 (Table 6.1).

**Algorithm 1** Diffusion-based graph augmentation with PC sampling

**Input:** Graph $G$ with node feature $\mathbf{X}$ and adjacency matrix $\mathbf{A}$, the denoising function for node feature $\mathbf{s_X}$ and adjacency matrix $\mathbf{s_A}$, the fine-tune loss $\mathcal{L}_{\mathbf{aug}}$, Lagevin MCMC step size $\beta$, scaling coefficient $\epsilon_1$
$\mathbf{A}^{(D)} \leftarrow \mathbf{A} + \mathbf{z}_A; \quad \mathbf{z}_A \sim \mathcal{N}(\mathbf{0}, \mathbf{I})$
$\mathbf{X}^{(D)} \leftarrow \mathbf{X} + \mathbf{z}_X; \quad \mathbf{z}_X \sim \mathcal{N}(\mathbf{0}, \mathbf{I})$
**for** $t = D - 1$ **to** 0 **do**
    $\hat{G}_{(t+1)} \sim p_{0t+1}(\hat{G}_{(t+1)}|G^{(t+1)})$     ▷ inner-loop sampling with another PC sampler
    $\mathbf{S}_A = \frac{1}{2}\mathbf{s_A}(G^{(t+1)}, t+1) - \frac{1}{2}\alpha \nabla_{\mathbf{A}^{(t)}} \mathcal{L}_{\mathbf{aug}}(\hat{G}_{(t+1)})$
    $\mathbf{S}_X = \frac{1}{2}\mathbf{s_X}(G^{(t+1)}, t+1) - \frac{1}{2}\alpha \nabla_{\mathbf{X}^{(t)}} \mathcal{L}_{\mathbf{aug}}(\hat{G}_{(t+1)})$
    $\tilde{\mathbf{A}}^{(t)} \leftarrow \mathbf{A}^{(t+1)} + g(t)^2 \mathbf{S}_A + g(t)\mathbf{z}_A; \quad \mathbf{z}_A \sim \mathcal{N}(\mathbf{0}, \mathbf{I})$     ▷ Predictor for adjacency matrix
    $\tilde{\mathbf{X}}^{(t)} \leftarrow \mathbf{X}^{(t+1)} + g(t)^2 \mathbf{S}_X + g(t)\mathbf{z}_X; \quad \mathbf{z}_X \sim \mathcal{N}(\mathbf{0}, \mathbf{I})$     ▷ Predictor for node features
    $\mathbf{A}^{(t)} \leftarrow \tilde{\mathbf{A}}^{(t)} + \frac{\beta}{2}\mathbf{S}_A + \epsilon_1 \sqrt{\beta}\mathbf{z}_A; \quad \mathbf{z}_A \sim \mathcal{N}(\mathbf{0}, \mathbf{I})$     ▷ Corrector for adjacency matrix
    $\mathbf{X}^{(t)} \leftarrow \tilde{\mathbf{X}}^{(t)} + \frac{\beta}{2}\mathbf{S}_X + \epsilon_1 \sqrt{\beta}\mathbf{z}_X; \quad \mathbf{z}_X \sim \mathcal{N}(\mathbf{0}, \mathbf{I})$     ▷ Corrector for node features
**end for**
return $G' = (\mathbf{A}^{(0)}, \mathbf{X}^{(0)})$

---

**Algorithm 2** The data-centric knowledge transfer framework: learning from unlabeled graphs

**Input:** Given unlabeled graphs from the space $\mathcal{G}^{[U]}$, randomly initialized score models $\mathbf{s_X}$ and $\mathbf{s_A}$ for node feature and graph adjacency matrix, respectively, the total diffusion time step $T$.
**while** $\mathbf{s_X}$ and $\mathbf{s_A}$ not converged **do**
    Sample $G = (\mathbf{X}, \mathbf{A}) \in \mathcal{G}^{[U]}$
    Sample $t \in \text{Uniform}(1, 2, \ldots, T)$
    Sample $\mathbf{z}_A \sim \mathcal{N}(\mathbf{0}, \mathbf{I})$
    Sample $\mathbf{z}_X \sim \mathcal{N}(\mathbf{0}, \mathbf{I})$
    Sample $\hat{G}$ with $t, \mathbf{z}_A, \mathbf{z}_X$ and Eq. (6.7)
    Optimize $\mathbf{s_A}$ with the gradient:
        $\nabla \|\mathbf{z}_A - \mathbf{s_A}(\hat{G}, t)\|^2$
    Optimize $\mathbf{s_X}$ with the gradient:
        $\nabla \|\mathbf{z}_X - \mathbf{s_X}(\hat{G}, t)\|^2$
**end while**

---

In this section, we present and analyze experimental results to demonstrate the outstanding performance of DCT, the usefulness of new optimization objectives, the effect of hyperparameters and iterative process, and the interpretability of "visible" knowledge transfer from unlabeled graphs.

## 6.4 Experiments

**Algorithm 3** The data-centric knowledge transfer framework: generating task-specific labeled graphs

---
**Input:** Given task $k$ with the graph-label space $(\mathcal{G}, \mathcal{Y})$, a randomly initialized prediction model $f_\theta$, the well-trained score model $\mathbf{s} = (\mathbf{s_X}, \mathbf{s_A})$, the training data set $\{G_i, y_i\}_i^{N_t}$, total training epoch $e$, the hyper-parameter $n$

**for** current epoch $e_i$ from 1 to $e$ **do**
    Train $f_\theta$ on current training data $\{G_i, y_i\}_i^{N_t}$
    **if** $e_i$ is divisible by the augmentation interval **then**
        Select $n$ graph-label pairs with the lowest training loss from $\{G_i, y_i\}_i^{N_t}$
        Get the augmented examples $\{G'_i, y'_i\}_i^n$ by Algorithm 1 with the selected examples
        Update $\{G_i, y_i\}_i^{N_t}$ with $\{G'_i, y'_i\}_i^n$, e.g., add $\{G'_i, y'_i\}_i^n$ to $\{G_i, y_i\}_i^{N_t}$.
    **end if**
**end for**

---

## 6.4 Experiments

### 6.4.1 Experimental Settings

Experiments are conducted on 15 datasets, including eight classification and seven regression tasks from chemistry, material science, and biology. We use Area under the ROC curve (AUC) for classification and mean absolute error (MAE) for regression.

**Table 6.1** Statistics of datasets for graph property prediction in different domains

| Data type | Dataset | # Graphs | Prediction task | # Task | Avg./max # nodes | Avg./max # edges |
|---|---|---|---|---|---|---|
| Molecules | ogbg-HIV | 41,127 | Classification | 1 | 25.5/222 | 54.9/502 |
| | ogbg-ToxCast | 8,576 | Classification | 617 | 18.8/124 | 38.5/268 |
| | ogbg-Tox21 | 7,831 | Classification | 12 | 18.6/132 | 38.6/290 |
| | IP | 2,039 | Classification | 1 | 24.1/132 | 51.9/290 |
| | ogbg-BACE | 1,513 | Classification | 1 | 34.1/97 | 73.7/202 |
| | ogbg-ClinTox | 1,477 | Classification | 2 | 26.2/136 | 55.8/286 |
| | ogbg-SIDER | 1,427 | Classification | 27 | 33.6/492 | 70.7/1010 |
| | Mol-Lipo | 4200 | Regression | 1 | 27/115 | 59/236 |
| | Mol-ESOL | 1128 | Regression | 1 | 13.3/55 | 27.4/124 |
| | Mol-FreeSolv | 642 | Regression | 1 | 8.7/24 | 16.8/50 |
| Polymers | GlassTemp | 7,174 | Regression | 1 | 36.7/166 | 79.3/362 |
| | MeltingTemp | 3,651 | Regression | 1 | 26.9/102 | 55.4/212 |
| | ThermCond | 759 | Regression | 1 | 21.3/71 | 42.3/162 |
| | $O_2$Perm | 595 | Regression | 1 | 37.3/103 | 82.1/234 |
| Proteins | PPI | 88000 | Classification | 40 | 49.4/111 | 890.8/11556 |

**Molecule Classification and Regression Tasks**: Seven molecule classification and three molecule regression tasks are from open graph benchmark (Hu et al. 2020). They were originally collected by MoleculeNet (Wu et al. 2018) and used to predict molecular properties. They include (1) inhibition to HIV virus replication in ogbg-HIV, (2) toxicological properties of 617 types in ogbg-ToxCast, (3) toxicity measurements such as nuclear receptors and stress response in ogbg-Tox21, (4) blood–brain barrier permeability in ogbg-BBBP, (5) inhibition to human $\beta$-secretase 1 in ogbg-BACE, (6) FDA approval status or failed clinical trial in ogbg-ClinTox, (7) having drug side effects of 27 system organ classes in ogbg-SIDER, (8) predicting the property of lipophilicity in Mol-Lipo, (9) predicting the water solubility (log solubility in mols per litre) from chemical structures in Mol-ESOL, (10) predicting the hydration free energy of molecules in water in Mol-FreeSolv. For all molecule datasets, we use the scaffold splitting procedure as the open graph benchmark adopted (Hu et al. 2020). It attempts to separate structurally different molecules into different subsets, which provides a more realistic estimate of model performance in experiments (Wu et al. 2018).

**Polymer Regression Tasks**: Four polymer regression tasks include GlassTemp, MeltingTemp, ThermCond, and $O_2$Perm. They are used to predict different polymer properties such as *glass transition temperature* (°C), *melting temperature* (°C), *thermal conductivity* (W/mK) and *oxygen permeability* (Barrer). GlassTemp and MeltingTemp are collected from PolyInfo, which is the largest web-based polymer database (Otsuka et al. 2011). The ThermCond dataset is from molecular dynamics simulation and is an extension from the dataset used in (Ma et al. 2022). The $O_2$Perm dataset is created from the Membrane Society of Australasia portal, consisting of various gas permeability data (Thornton et al. 2012). Since a polymer is built from repeated units, researchers often use a single unit graph with polymerization points as polymer graphs to predict properties. Different from molecular graphs, two polymerization points are two special nodes (see "∗" in Fig. 6.3), indicating the polymerization of monomers (Cormack and Elorza 2004). For all the polymer tasks, we randomly split by 60%/10%/30% for training, validation, and test.

**Protein Classification Task**: The protein function prediction using protein-protein interaction graphs (Hu et al. 2019). A node is a protein without attributes, an edge is a relation type between two proteins such as co-expression and co-occurrence. In DCT, we treat all the relations as the undirected edge without attributes.

**Baselines and Implementation**: GIN's hyper-parameters are tuned for different tasks with an early stop on the validation set. We generally implement pre-training baselines following their own settings. For molecule and polymer property prediction and protein function prediction, the pre-trained GIN models with self-supervised tasks such as EDGEPRED, ATTRMASK, CONTEXTPRED in (Hu et al. 2019), INFOMAX (Velickovic et al. 2019) are available. So we directly use them. For other self-supervised methods, we implement their codes with default hyper-parameters. Following their settings, we use 2M ZINC15 (Sterling and Irwin 2015) to pre-train GIN models for molecule and polymer property prediction. We use 306K unlabeled protein-protein interaction ego-networks (Hu et al. 2019) to pre-train the GIN for the downstream protein function property prediction. For self-training with real unlabeled graphs and INFOGRAPH (Sun et al. 2020), we use 113K QM9 (Ramakrishnan et al. 2014). For self-training with generated unlabeled graphs, we train the diffusion model (Jo et al. 2022) on the real QM9 dataset and then produce the same number of generated unlabeled graphs. The QM9 (Ramakrishnan et al. 2014) is also used to train the diffusion model in DCT.

### 6.4.2 RQ1: Outstanding Property Prediction Performance

We report the model performance using mean and standard deviation over 10 runs Table 6.2. DCT is the best on all 15 tasks compared to the state-of-the-art baselines. The observations are:

(1) **GIN is the most competitive baseline and outperforms self-supervised learning methods**. On 7 of 15 tasks, GIN outperforms all the 7 self-supervised learning methods. Because self-supervised pre-training imposes constraints on the model architecture, it undermines the true power of GNNs and under-performs the GNNs that are properly used.

(2) **Self-training and GDA methods perform better than GIN but cannot effectively learn from unlabeled data**. Self-training (ST- REAL and ST- GEN) is often the best baseline in regression tasks. GDA (GREA and G- MIXUP) methods outperform self-training in most classification tasks except ogbg-SIDER, because they are often designed to exploit categorical labeled data and remain under-explored for regression. Although self-training benefits from selecting unlabeled examples in some graph regression tasks, they are *negatively* affected by the unlabeled graphs in the classification tasks such as ogbg-ToxCast and ogbg-ClinTox. As indicated in Fig. 6.4, it is inappropriate to pseudo-label unlabeled graphs in self-training due to the huge gap between the unlabeled data and target task.

(3) **DCT transfers useful knowledge from unlabeled data by data augmentation**. DCT outperforms the best baseline relatively by +3.9%, +13.4%, and +10.2% when there are only 1,210, 513, and 4,303 training graphs on ogbg-BACE, Mol-FreeSolv, and GlassTemp, respectively. Compared to the self-supervised baselines, the improvement from DCT is more significant, so the knowledge transfer is more effective. For example, on Mol-FreeSolv and $O_2$Perm, DCT performs better than the best self-supervised baselines relatively by +45.8% and +8.0%, respectively. On regression tasks that involve knowledge transfer across domains (*e.g.*, from molecules to polymers), DCT reduces MAE relatively by 1.9% $\sim$ 10.2% compared to the best baseline. All these results demonstrate the outstanding performance of task-specific data augmentation in DCT.

**Table 6.2** Mean(Std) on tasks from different fields

| # Training graphs | | Molecule classification: AUC (%) ↑ | | | | | | |
|---|---|---|---|---|---|---|---|---|
| | | ogbg-HIV 32,901 | ogbg-ToxCast 6,860 | ogbg-Tox21 6,264 | IP1,631 | ogbg-BACE 1,210 | ogbg-ClinTox 1,181 | ogbg-SIDER 1,141 |
| GIN | | 77.4(1.2) | 66.9(0.2) | 76.0(0.6) | 67.5(2.7) | 77.5(2.8) | 88.8(3.8) | 58.1(0.9) |
| Self-supervised | EDGEPRED | 78.1(1.3) | 63.9(0.4) | 75.5(0.4) | 69.9(0.5) | 79.5(1.0) | 62.9(2.3) | 59.7(0.8) |
| | ATTRMASK | 77.1(1.7) | 64.2(0.5) | 76.6(0.4) | 63.9(1.2) | 79.3(0.7) | 70.4(1.1) | 60.7(0.4) |
| | CONTEXTPRED | 78.4(0.1) | 63.7(0.3) | 75.0(0.1) | 68.8(1.6) | 75.7(1.0) | 63.2(6.5) | 60.7(0.8) |
| | INFOMAX | 75.4(1.8) | 61.7(1.0) | 75.5(0.4) | 69.2(0.5) | 76.8(0.2) | 73.0(0.2) | 58.6(0.5) |
| | JOAO | 76.2(0.2) | 64.8(0.3) | 74.8(0.5) | 69.3(2.5) | 75.9(3.9) | 69.4(4.5) | 60.8(0.6) |
| | GRAPHLOG | 74.8(1.1) | 63.2(0.8) | 75.4(0.8) | 67.5(2.3) | 80.4(3.6) | 69.0(6.6) | 57.0(0.9) |
| | MGSSL | 77.1(1.1) | 65.7(0.4) | 77.2(0.3) | 66.9(0.9) | 81.3(2.4) | 69.8(5.0) | 63.6(1.0) |
| | D-SLA | 76.9(0.9) | 60.8(1.2) | 76.1(0.1) | 62.6(1.0) | 80.3(0.6) | 78.3(2.4) | 55.1(1.0) |
| Semi-SL | INFOGRAPH | 73.3(0.7) | 61.5(1.1) | 67.6(0.9) | 61.6(4.4) | 75.9(1.8) | 62.2(5.5) | 56.3(2.3) |
| | ST-REAL | 78.3(0.6) | 64.5(1.0) | 76.2(0.5) | 66.7(1.9) | 77.4(1.8) | 82.2(2.4) | 60.8(1.2) |
| | ST-GEN | 77.9(1.6) | 65.1(1.0) | 75.8(0.9) | 66.3(1.5) | 78.4(3.0) | 87.3(1.3) | 59.3(1.3) |
| GDA | FLAG | 74.6(1.7) | 59.9(1.6) | 76.9(0.7) | 66.6(1.0) | 79.1(1.2) | 85.1(3.4) | 57.6(2.3) |
| | GREA | 79.3(0.9) | 67.5(0.7) | 77.2(1.2) | 69.7(1.3) | 82.4(2.4) | 87.9(3.7) | 60.1(2.0) |
| | G-MIXUP | 77.1(1.1) | 55.6(1.1) | 64.6(0.4) | 70.2(1.0) | 77.8(3.3) | 60.2(7.5) | 56.8(3.5) |
| | DCT (Ours) | **79.5**(1.0) | **68.1**(0.2) | **78.2**(0.2) | **70.8**(0.5) | **85.6**(0.6) | **92.1**(0.8) | **63.9**(0.3) |

(continued)

## 6.4 Experiments

**Table 6.2** (continued)

| # Training graphs | | Molecule regression: MAE ↓ | | | Polymer regression: MAE ↓ | | | | Bio: AUC (%) ↑ |
|---|---|---|---|---|---|---|---|---|---|
| | | Mol-Lipo 3,360 | Mol-ESOL 902 | Mol-FreeSolv 513 | GlassTemp 4,303 | MeltingTemp 2,189 | ThermCond 455 | O₂Perm 356 | PPI 60,715 |
| GIN | | 0.545(0.019) | 0.766(0.016) | 1.639(0.146) | 26.4(0.2) | 40.9(2.2) | 3.25(0.19) | 201.3(45.0) | 69.1(0.0) |
| Self-supervised | EDGEPRED | 0.585(0.008) | 1.062(0.066) | 2.249(0.150) | 27.6(1.4) | 47.4(2.8) | 3.69(0.50) | 207.3(41.7) | 63.7(1.1) |
| | ATTRMASK | 0.573(0.009) | 1.041(0.041) | 1.952(0.088) | 27.7(0.8) | 45.8(2.6) | 3.17(0.32) | 179.9(30.8) | 64.1(1.8) |
| | CONTEXTPRED | 0.592(0.007) | 0.971(0.027) | 2.193(0.151) | 27.6(0.3) | 46.7(1.9) | 3.15(0.24) | 191.2(35.2) | 62.0(1.2) |
| | INFOMAX | 0.581(0.009) | 0.935(0.018) | 2.197(0.129) | 27.5(0.8) | 46.5(2.8) | 3.31(0.25) | 231.0(52.6) | 63.3(1.2) |
| | JOAO | 0.596(0.016) | 1.098(0.037) | 2.465(0.095) | 27.5(0.2) | 46.0(0.2) | 3.55(0.26) | 207.7(43.7) | 61.5(1.2) |
| | GRAPHLOG | 0.577(0.010) | 1.109(0.059) | 2.373(0.283) | 29.5(1.3) | 50.3(3.3) | 3.01(0.17) | 229.7(48.3) | 62.1(0.6) |
| | MGSSL | 0.569(0.007) | 0.998(0.031) | 1.956(0.077) | 26.9(0.4) | 42.7(1.2) | 3.10(0.14) | 201.1(31.9) | N.A. |
| | D-SLA | 0.563(0.004) | 1.064(0.030) | 2.190(0.149) | 27.5(1.0) | 51.7(2.5) | 2.71(0.08) | 257.8(30.2) | 65.0(1.2) |
| Semi-SL | INFOGRAPH | 0.793(0.094) | 1.285(0.093) | 3.710(0.418) | 30.8(1.2) | 51.2(5.1) | 2.75(0.15) | 207.2(21.8) | 67.7(0.4) |
| | ST-REAL | 0.526(0.009) | 0.788(0.070) | 1.770(0.251) | 26.6(0.3) | 42.3(1.2) | 2.64(0.07) | 256.0(17.5) | 68.9(0.1) |
| | ST-GEN | 0.531(0.031) | 0.724(0.082) | 1.547(0.082) | 26.8(0.3) | 42.0(0.9) | 2.70(0.03) | 262.2(10.1) | 68.6(0.6) |
| GDA | FLAG | 0.528(0.012) | 0.755(0.039) | 1.565(0.098) | 26.6(1.3) | 44.2(2.0) | 3.05(0.10) | 177.7(60.7) | 69.2(0.2) |
| | GREA | 0.586(0.036) | 0.805(0.135) | 1.829(0.368) | 26.7(1.0) | 41.1(0.8) | 3.23(0.18) | 194.0(45.5) | 68.8(0.2) |
| | DCT (Ours) | **0.516**(0.071) | **0.717**(0.020) | **1.339**(0.075) | **23.7**(0.2) | **38.0**(0.8) | **2.59**(0.11) | **165.6**(24.3) | **69.5**(0.2) |

The best mean is **bold**. The best baseline is underlined. Results are highlighted if unlabeled graphs bring significant negative impacts compared to GIN. The MAE for ThermCond is scaled × 100. G-MIXUP was proposed for classification. MGSSL was proposed for molecules

**Table 6.3** Comprehensive ablation studies for DCT on tasks ogbg-BACE, ogbg-SIDER, Mol-FreeSolv, and O$_2$Perm. Objectives include minimizing $\mathcal{I}_1(G', G)$ and/or maximizing $\mathcal{I}_2(G', y)$

| | | Objectives | | Classification | | Regression | |
|---|---|---|---|---|---|---|---|
| | | $\mathcal{I}_1(G', G)$ | $\mathcal{I}_2(G', y)$ | BACE | SIDER | FreeSolv | O$_2$Perm |
| Top baseline method | | | | 82.4(2.4) | 60.8(1.2) | 1.547(0.082) | 177.7(60.7) |
| Unlabeled data sources | QM9 | ✗ | ✗ | 84.4(2.6) | 63.7(0.3) | 1.473(0.192) | 177.4(27.3) |
| | | ✓ | ✗ | 85.2(1.3) | 63.7(0.2) | 1.415(0.145) | 171.4(14.0) |
| | | ✗ | ✓ | 84.7(1.8) | 63.8(0.5) | 1.344(0.096) | 172.6(32.9) |
| | | ✓ | ✓ | **85.6**(0.6) | **63.9**(0.3) | **1.339**(0.075) | **165.6**(24.3) |
| | ZINC | ✗ | ✗ | 82.8(1.8) | 63.5(0.7) | 1.524(0.219) | 175.5(11.9) |
| | | ✓ | ✗ | 83.3(2.2) | 63.5(0.7) | 1.455(0.207) | 172.4(60.8) |
| | | ✗ | ✓ | 84.3(0.6) | 63.5(0.6) | 1.514(0.214) | 171.5(26.0) |
| | | ✓ | ✓ | 84.9(0.4) | 63.7(0.7) | 1.408(0.092) | 169.3(15.3) |

### 6.4.3 RQ2: Ablation Studies and Performance Analysis

**Comprehensive ablation studies**: In Table 6.3, we investigate how the task-related objectives in Eq. (6.8) impact the performance of DCT. First, DCT outperforms the top baseline even if the two task-related optimization objectives are disabled. This is because DCT generates new training examples based on original labeled graphs: the data augmentation has already improved the diversity of the training dataset a little bit. Second, adding the objective $\mathcal{I}_1$ further improves the performance by encouraging the generation of diverse examples, because it minimizes the similarity between the original graph and augmented graph in the graph space. Third, we receive the best performance of DCT when it combines $\mathcal{I}_1$ and $\mathcal{I}_2$ objectives to generate task-related and diverse augmented graphs. When we change the unlabeled data source from QM9 to the ZINC dataset from (Jo et al. 2022), similar observations confirm the necessity of the task-related objectives.

**Effect of hyper-parameters**: The impacts of three hyper-parameters of DCT are studied: the number of perturbation steps $D$, the number of negative samples $M$ in Eq. (6.9), and the number of augmented graphs in each iteration (*i.e.,* top-$n$ % selected graph for augmentation). Results from Fig. 6.5 show that DCT is robust to a wide range of $D$ and $M$ valued from 0 to 10. They suggest that $D$ and $M$ can be set as 5 in most cases. As for the number of the augmented graphs in each iteration, results show that noisy graphs are often created when $n$ is higher than 30%, because the predictor cannot effectively guide the data augmentation for those labeled graphs whose labels are hard to predict. So, 10% is suggested as the default of top-$n$%.

## 6.4 Experiments

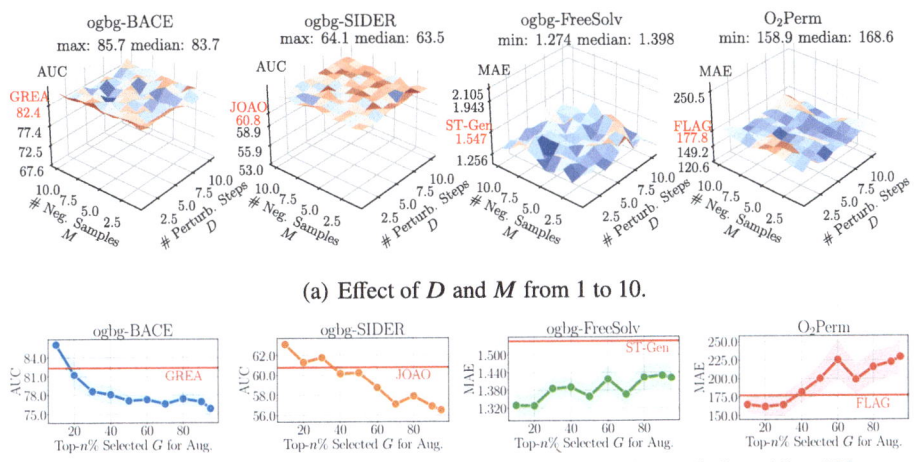

(a) Effect of $D$ and $M$ from 1 to 10.

(b) Effect of top-$n$% labeled graphs for data augmentation, where $n$ is from 10 to 100.

**Fig. 6.5** Effect of hyper-parameters, including the number of perturbation steps $D \in [1, 10]$, the number of negative graphs $M \in [1, 10]$, and top-$n$% labeled graphs whose labels are predicted the most accurately and that are selected for data augmentation, where $n \in [10, 100]$

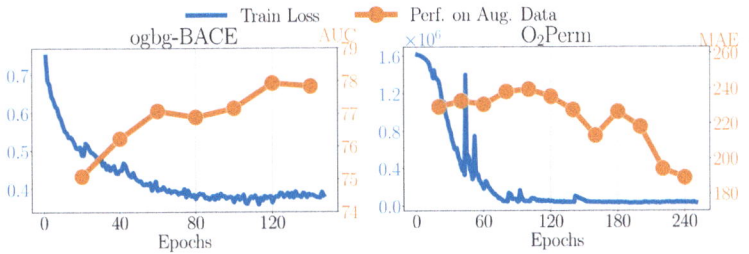

**Fig. 6.6** Data augmentation and model training mutually enhance each other over epochs. The predictor is saved every 20 epochs to guide the generation of augmented graphs. The performance of GIN trained on these augmented graphs reflects the quality of the augmented data

**Iterative process**: Figure 6.6 investigates the relationship between the quality of augmented graphs and the accuracy of property prediction models. We save a predictor checkpoint every 20 epochs to guide the generation of the augmented examples. We evaluate the quality of augmented graphs by using them to train GIN and report AUC/MAE. The data augmentation gradually decreases the training loss of property prediction. On the other hand, the increased GIN performance indicates that the quality of augmented examples is also improved over epochs. The data augmentation and predictor training mutual enhance each other.

### 6.4.4 RQ3: Case Study for the Interpretability of Visible Knowledge Transfer

Knowledge transfer by data augmentation gives visible examples, allowing us to study what is learned. We visualize a few augmented graphs in DCT using ogbg-BACE and $O_2$Perm. We adapt top-k pooling (Knyazev et al. 2019) to select the subgraphs that GIN used for prediction. The selected subgraphs are highlighted in green in Fig. 6.7. The three examples show that *the augmented graphs can identify and preserve the core structures* that GIN uses to predict property values. We have more observations:

(1) These augmented graphs are chemically valid, showing that *concepts such as some chemical rules from the unlabeled graphs are successfully transferred to downstream tasks*. To further validate this point, we gathered 1,000 task-specific graphs generated in the intermediate steps on the tasks of ogbg-BACE, ogbg-BBBP, Mol-FreeSolv, and $O_2$Perm. We then assessed the chemical validity of these graphs and observed that the validity is 92.8%, 87.9%, 97.4%, and 62.1%, respectively. Results show that transferring knowledge from pre-trained molecular data to target molecules yields relatively high chemical validity. However, the validity drops to 62% when transferring knowledge from pre-trained molecular data to target polymer data. This finding indicates that the transferability of chemical rules becomes more challenging when the distribution gap between the pre-training data and downstream task data is larger.

(2) Regarding task-specific knowledge, it is known that the fluorine atom and the methyl group are usually negatively and positively correlated to the permeability, respec-

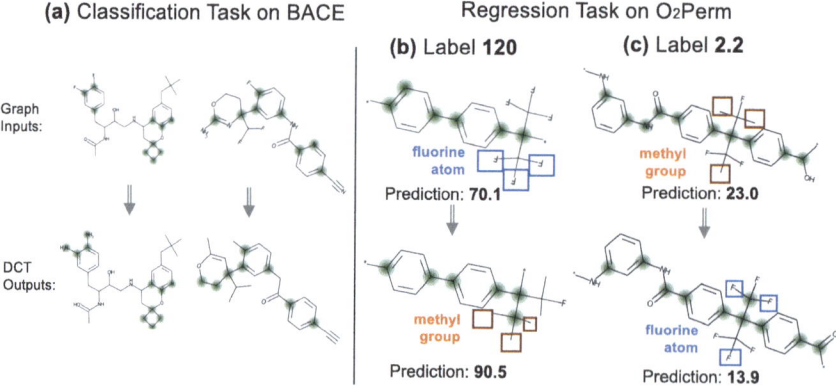

**Fig. 6.7** Case studies of augmented graphs. The green highlighted subgraphs are from GIN with top-k pooling. Examples show that the augmented graphs from DCT preserve the core structures of original graphs. Key concepts in the unlabeled graphs like chemical validity are transferred to downstream tasks. Domain knowledge such as the relationship between the permeability and the fluorine atom/methyl group is captured to guide task-specific generation

tively (Park et al. 2003; Corrado and Guo 2020). The augmented examples show that *DCT captures this domain knowledge with the task-related objectives*. In example (b), DCT replaces most of the fluorine atoms with the methyl groups. It encourages GIN to learn the positive relationship between the methyl group and the permeability so that GIN predicts a high label value. In example (c), DCT replaces the methyl groups with fluorine atoms. It encourages GIN to learn the negative relationship between the fluorine atom and the permeability so that GIN predicts a low label value.

## 6.5 Conclusion

In this chapter, we introduced an attempt to transfer minimal sufficient knowledge from unlabeled graphs by data augmentation. We presented a data-centric framework to use the diffusion model trained on the unlabeled graphs and use two task-related objectives to generate task-specific augmented graphs. Experiments demonstrated the performance of the framework through visible augmented examples. It is better than self-supervised learning, self-training, and graph data augmentation methods on as many as 15 tasks.

## References

H.-J. Böhm, A. Flohr, and M. Stahl. Scaffold hopping. *Drug discovery today: Technologies*, 1(3):217–224, 2004.

N. Brown. Bioisosteres and scaffold hopping in medicinal chemistry. *Molecular informatics*, 33(6-7):458–462, 2014.

T. Brown, B. Mann, N. Ryder, M. Subbiah, J. D. Kaplan, P. Dhariwal, A. Neelakantan, P. Shyam, G. Sastry, A. Askell, et al. Language models are few-shot learners. *Advances in neural information processing systems*, 33:1877–1901, 2020.

P. A. Cormack and A. Z. Elorza. Molecularly imprinted polymers: synthesis and characterisation. *Journal of chromatography B*, 804(1):173–182, 2004.

T. Corrado and R. Guo. Macromolecular design strategies toward tailoring free volume in glassy polymers for high performance gas separation membranes. *Molecular Systems Design & Engineering*, 5(1):22–48, 2020.

P. Dhariwal and A. Nichol. Diffusion models beat gans on image synthesis. *Advances in Neural Information Processing Systems*, 34:8780–8794, 2021.

X. Han, Z. Jiang, N. Liu, and X. Hu. G-mixup: Graph data augmentation for graph classification. *arXiv preprint* arXiv:2202.07179, 2022.

K. He, X. Chen, S. Xie, Y. Li, P. Dollár, and R. Girshick. Masked autoencoders are scalable vision learners. In *Proceedings of the IEEE/CVF Conference on Computer Vision and Pattern Recognition*, pages 16000–16009, 2022.

W. Hu, B. Liu, J. Gomes, M. Zitnik, P. Liang, V. Pande, and J. Leskovec. Strategies for pre-training graph neural networks. *arXiv preprint* arXiv:1905.12265, 2019.

W. Hu, M. Fey, M. Zitnik, Y. Dong, H. Ren, B. Liu, M. Catasta, and J. Leskovec. Open graph benchmark: Datasets for machine learning on graphs. *Neural Information Processing Systems (NeurIPS)*, 2020.

K. Huang, T. Fu, W. Gao, Y. Zhao, Y. H. Roohani, J. Leskovec, C. W. Coley, C. Xiao, J. Sun, and M. Zitnik. Therapeutics data commons: Machine learning datasets and tasks for drug discovery and development. In *Thirty-fifth Conference on Neural Information Processing Systems Datasets and Benchmarks Track*, 2021.

J. Jo, S. Lee, and S. J. Hwang. Score-based generative modeling of graphs via the system of stochastic differential equations. In *International Conference on Machine Learning*, volume 162, pages 10362–10383. PMLR, 2022.

D. Kim, J. Baek, and S. J. Hwang. Graph self-supervised learning with accurate discrepancy learning. *Advances in Neural Information Processing Systems*, 2022.

B. Knyazev, G. W. Taylor, and M. Amer. Understanding attention and generalization in graph neural networks. *Advances in neural information processing systems*, 32, 2019.

K. Kong, G. Li, M. Ding, Z. Wu, C. Zhu, B. Ghanem, G. Taylor, and T. Goldstein. Robust optimization as data augmentation for large-scale graphs. In *Proceedings of the IEEE/CVF Conference on Computer Vision and Pattern Recognition*, pages 60–69, 2022.

G. Liu, T. Zhao, J. Xu, T. Luo, and M. Jiang. Graph rationalization with environment-based augmentations. In *Proceedings of the 28th ACM SIGKDD Conference on Knowledge Discovery and Data Mining*, pages 1069–1078, 2022.

Y. Luo, M. McThrow, W. Y. Au, T. Komikado, K. Uchino, K. Maruhash, and S. Ji. Automated data augmentations for graph classification. *arXiv preprint* arXiv:2202.13248, 2022.

R. Ma, H. Zhang, J. Xu, L. Sun, Y. Hayashi, R. Yoshida, J. Shiomi, J.-x. Wang, and T. Luo. Machine learning-assisted exploration of thermally conductive polymers based on high-throughput molecular dynamics simulations. *Materials Today Physics*, 28:100850, 2022.

Ł. Maziarka, A. Pocha, J. Kaczmarczyk, K. Rataj, T. Danel, and M. Warchoł. Mol-cyclegan: a generative model for molecular optimization. *Journal of Cheminformatics*, 12(1):1–18, 2020.

A. v. d. Oord, Y. Li, and O. Vinyals. Representation learning with contrastive predictive coding. *arXiv preprint* arXiv:1807.03748, 2018.

S. Otsuka, I. Kuwajima, J. Hosoya, Y. Xu, and M. Yamazaki. Polyinfo: Polymer database for polymeric materials design. In *2011 International Conference on Emerging Intelligent Data and Web Technologies*, pages 22–29. IEEE, 2011.

S.-H. Park, K.-J. Kim, W.-W. So, S.-J. Moon, and S.-B. Lee. Gas separation properties of 6fda-based polyimide membranes with a polar group. *Macromolecular Research*, 11:157–162, 2003.

B. Poole, S. Ozair, A. Van Den Oord, A. Alemi, and G. Tucker. On variational bounds of mutual information. In *International Conference on Machine Learning*, pages 5171–5180. PMLR, 2019.

R. Ramakrishnan, P. O. Dral, M. Rupp, and O. A. Von Lilienfeld. Quantum chemistry structures and properties of 134 kilo molecules. *Scientific data*, 1(1):1–7, 2014.

Y. Rong, W. Huang, T. Xu, and J. Huang. Dropedge: Towards deep graph convolutional networks on node classification. *arXiv preprint* arXiv:1907.10903, 2019.

Y. Rong, Y. Bian, T. Xu, W. Xie, Y. Wei, W. Huang, and J. Huang. Self-supervised graph transformer on large-scale molecular data. *Advances in Neural Information Processing Systems*, 33:12559–12571, 2020.

S. Soatto and A. Chiuso. Visual representations: Defining properties and deep approximations. *International Conference on Learning Representations*, 2016.

Y. Song, J. Sohl-Dickstein, D. P. Kingma, A. Kumar, S. Ermon, and B. Poole. Score-based generative modeling through stochastic differential equations. *International Conference on Learning Representations*, 2021.

# References

T. Sterling and J. J. Irwin. Zinc 15–ligand discovery for everyone. *Journal of chemical information and modeling*, 55(11):2324–2337, 2015.

F.-Y. Sun, J. Hoffmann, V. Verma, and J. Tang. Infograph: Unsupervised and semi-supervised graph-level representation learning via mutual information maximization. *International Conference on Learning Representations*, 2020.

M. Sun, J. Xing, H. Wang, B. Chen, and J. Zhou. Mocl: data-driven molecular fingerprint via knowledge-aware contrastive learning from molecular graph. In *Proceedings of the 27th ACM SIGKDD Conference on Knowledge Discovery & Data Mining*, pages 3585–3594, 2021.

R. Sun, H. Dai, and A. W. Yu. Does gnn pretraining help molecular representation? *Advances in Neural Information Processing Systems*, 2022.

A. Thornton, L. Robeson, B. Freeman, and D. Uhlmann. Polymer gas separation membrane database, 2012. URL https://research.csiro.au/virtualscreening/membrane-database-polymer-gas-separation-membranes/.

Y. Tian, C. Sun, B. Poole, D. Krishnan, C. Schmid, and P. Isola. What makes for good views for contrastive learning? *Advances in Neural Information Processing Systems*, 33:6827–6839, 2020.

P. Trivedi, E. S. Lubana, M. Heimann, D. Koutra, and J. J. Thiagarajan. Analyzing data-centric properties for graph contrastive learning. In *Advances in Neural Information Processing Systems*, 2022.

P. Velickovic, W. Fedus, W. L. Hamilton, P. Liò, Y. Bengio, and R. D. Hjelm. Deep graph infomax. *ICLR (Poster)*, 2(3):4, 2019.

Z. Wu, B. Ramsundar, E. N. Feinberg, J. Gomes, C. Geniesse, A. S. Pappu, K. Leswing, and V. Pande. Moleculenet: a benchmark for molecular machine learning. *Chemical science*, 9(2):513–530, 2018.

K. Xu, W. Hu, J. Leskovec, and S. Jegelka. How powerful are graph neural networks? *arXiv preprint* arXiv:1810.00826, 2018.

Y. You, T. Chen, Y. Shen, and Z. Wang. Graph contrastive learning automated. In *International Conference on Machine Learning*, pages 12121–12132. PMLR, 2021.

Z. Zhang, Q. Liu, H. Wang, C. Lu, and C.-K. Lee. Motif-based graph self-supervised learning for molecular property prediction. *Advances in Neural Information Processing Systems*, 34:15870–15882, 2021.

The manufacturer's authorised representative in the EU is Springer Nature Customer Service Centre GmbH, Europaplatz 3, 69115 Heidelberg, Germany. If you have any concerns regarding our products, please contact ProductSafety@springernature.com

Printed and bound by CPI Group (UK) Ltd, Croydon, CR0 4YY

26/03/2026

02078939-0019